T0135580

Feeding ecology of and lead exposure in a top predator: the white-tailed eagle (*Haliaeetus albicilla*)

Dissertation zur Erlangung des akademischen Grades des Doktors
der Naturwissenschaften (Dr. rer. nat.)

Eingereicht im Fachbereich Biologie, Chemie, Pharmazie
der Freien Universität Berlin

vorgelegt von
Mirjam Nadjafzadeh
aus Dinslaken

Berlin, Januar 2011

Bibliografische Information der Deutschen Nationalbibliothek

Die Deutsche Nationalbibliothek verzeichnet diese Publikation in der Deutschen Nationalbibliografie; detaillierte bibliografische Daten sind im Internet über http://dnb.d-nb.de abrufbar.

©Copyright Logos Verlag Berlin GmbH 2011
Alle Rechte vorbehalten.

ISBN 978-3-8325-2989-5

Logos Verlag Berlin GmbH
Comeniushof, Gubener Str. 47,
10243 Berlin
Tel.: +49 (0)30 42 85 10 90
Fax: +49 (0)30 42 85 10 92
INTERNET: http://www.logos-verlag.de

Diese Dissertation wurde am Leibniz-Institut für Zoo- und Wildtierforschung Berlin (Direktor: Prof. Dr. Heribert Hofer) im Zeitraum Juli 2006 bis Januar 2011 angefertigt und am Institut für Biologie des Fachbereichs Biologie, Chemie, Pharmazie der Freien Universität Berlin eingereicht.

1. Gutachter: Prof. Dr. Heribert Hofer

2. Gutachter: Prof. Dr. Silke Kipper

Disputation am 29. März 2011

This thesis is based on the following manuscripts:

1. Nadjafzadeh, M., Hofer, H. & Krone, O. (submitted). The link between feeding ecology and lead poisoning in white-tailed eagles.

2. Nadjafzadeh, M., Hofer, H. & Krone, O. (submitted). Sit-and-wait for large prey: foraging strategy and prey choice of white-tailed eagles.

3. Nadjafzadeh, M., Voigt, C. & Krone, O. (submitted). Specialisation in a generalist predator? Evidence from stable isotope analysis for spatial, seasonal and individual variation in diet composition of white-tailed eagles.

4. Nadjafzadeh, M., Hofer, H. & Krone, O. (submitted). Who ingests what and how much? An experimental approach to simulate lead exposure and food processing of white-tailed eagles and other scavengers at shot mammalian carcasses.

CONTENT

Chapter 1

General introduction and thesis outline 1

Chapter 2

The link between feeding ecology and lead poisoning in white-tailed eagles 16

Chapter 3

Sit-and-wait for large prey: foraging strategy and prey choice of white-tailed eagles 41

Chapter 4

Specialisation in a generalist predator? Evidence from stable isotope analysis for spatial, seasonal and individual variation in diet composition of white-tailed eagles 67

Chapter 5

Who ingests what and how much? An experimental approach to simulate lead exposure and food processing of white-tailed eagles and other scavengers at shot mammalian carcasses 91

Chapter 6

General discussion 117

Summary 128

Zusammenfassung 130

Danksagung 133

CHAPTER 1

General introduction and thesis outline

Feeding ecology – a central issue in individual fitness and biological conservation

Natural selection favours traits that are useful in the struggle for survival and reproduction and thus individuals with the highest fitness (Darwin 1859). The fitness of an individual depends on the success of behaviours such as the search for mates, predator avoidance and food acquisition (Schoner 1971). Accordingly, an individual's foraging strategy and food choice should have a great influence on an animal's fitness. The theory of optimal foraging postulates that as a consequence of evolutionary selection pressures, animals maximise efficiency (profitability) in harvesting food to improve their fitness (Pyke *et al.* 1977; Stephens & Krebs 1986). In which way animals optimise their food intake can be examined by optimality models which are based on cost-benefit analyses. Since the costs and benefits of different foraging options are difficult or impossible to quantify in increments of fitness, proxy currencies (presumed to be correlates of fitness) are employed. Costs are commonly regarded as the energy expenditure for food acquisition, which in general is expressed as the time required to search for and handle a food item (Stephens & Krebs 1986). The individual's net amount of energy gained during the time spent searching for and handling food is considered as benefit. Profitability is then the difference between costs and benefits (Stephens & Krebs 1986).

One key optimal foraging model that has received attention from many scientists refers to food choice and assumes that the individual's pattern of choice of food type (i.e., its diet) will be such that the net rate of energy intake is maximised (Emlen 1966; McArthur & Pianka 1966; Pyke *et al.* 1977; Krebs 1978). The prerequisite for this model is that foragers should be able to distinguish between items of differing profitability and to select more profitable food types. Its main predictions are that (1) when food is abundant, only the most profitable food type should be eaten and there should be no partial preferences, (2) inclusion of other food types in the diet should not depend on their own abundance but on the abundance of more profitable types, and (3) a general decline in prey abundance should result in a wider diet niche. Hence, food availability, search effort and handling time play a key role in shaping foraging strategies of free-ranging animals.

Detailed information about the feeding ecology of a particular species is not only crucial to understand the interaction between an animal and its environment but also to develop conservation measures (Fraser & Gordon 1997; Korine *et al.* 1999; Sinclair *et al.* 2006). Since a sufficient supply of food is important for animal survival and reproduction, knowledge about the food supply in the range of rare or endangered species is essential to design action plans that result in successful and sustainable management (González *et al.* 2006; López-Bao *et al.* 2008). The examination of predator-prey relationships allows to determine the impact of predation on the prey populations (Côté & Sutherland 1997; Williams *et al.* 2004). Dietary investigations are necessary to assess the exposure of animals to poisonous substances in their habitats as they may have a substantial influence on population growth (Frey 1996; Rattner 2009). In particular, anthropogenic pollution caused by pesticides (Sibly *et al.* 2000; Helander *et al.* 2008) and toxic heavy metals (Scheuhammer 1987; Kenntner *et al.* 2001; Scheuhammer *et al.* 2007) constitutes a major hazard to wildlife.

Lead poisoning – a challenge for raptor conservation

Lead is a highly toxic heavy metal with profound effects on animal health and reproduction. It occurs naturally in parts of the environment but is now ubiquitous as a consequence of human activities (Pain *et al.* 1995). One widespread source of lead has over the years resulted in considerable avian mortality, namely lead from spent ammunition such as lead shot and lead-based bullets (Scheuhammer & Templeton 1998; Fisher *et al.* 2006; Mateo *et al.* 2007).

Lead acts as a non-specific poison affecting all body systems. Typical symptoms of lead toxicosis in birds are weight loss, anaemia, diarrhoea, and paralysis of the crop, gizzard, legs or wings (Lumeij 1985). The singular ingestion of a larger number of lead shot or lead bullet fragments (ten or more) can induce acute lead poisoning and birds usually die within a few days (Scheuhammer & Norris 1996). Victims of acute poisoning can appear to be in good condition, without obvious weight loss. Chronic lead poisoning follows the ingestion of a smaller number of lead shot or lead fragments. In these cases, signs of lead poisoning appear more gradually and affected birds die several weeks after ingesting the lead particles, often in a very emaciated condition (Sanderson & Bellrose 1986).

The two main groups of birds affected by lead poisoning are waterfowl and raptors (Locke & Thomas 1996; Fisher *et al.* 2006). Lead exposure in waterfowl through the ingestion of spent lead shot has been well documented worldwide (e.g., Bellrose 1959; Pain 1990; Mateo *et al.* 2001; Mateo *et al.* 2007). Waterfowl contract primary lead poisoning usually by the ingestion of lead shot along with or mistakenly for grit (Figuerola *et al.* 2005). In several countries, legislation exists to combat lead toxicosis in wetlands and/or in waterbirds (Boere 1995; Thomas & Owen 1996; Beintema 2001; Thomas & Guitart 2005). Less well examined, although of increasing concern, is lead exposure in raptors, currently reported in 35 species from 18 countries (Miller *et al.* 2002). Here, secondary lead poisoning occurs when raptors feeding either on wildlife killed by lead ammunition or waterfowl with lead shot pellets in their gastrointestinal tract (Pain *et al.* 1995; Scheuhammer & Templeton 1998; Martina *et al.* 2008). It has been suggested that carcasses of shot game mammals contaminated with lead bullet fragments provide a particular high exposure to lead and poisoning probably occurs in any scavenging raptor species (Locke & Thomas 1996; Kim *et al.* 1999; Meretsky *et al.* 2000; Hunt *et al.* 2006; Krone *et al.* 2009). However, little action has been taken against the use of lead-based bullets (Fisher *et al.* 2006). Two exceptions are in California (USA) and Hokkaido (Japan) where the use of lead-based bullets is officially banned to protect the critically endangered California condor (*Gymnogyps californianus*) and the vulnerable Steller's sea eagle (*Haliaeetus pelagicus*; Sieg *et al.* 2008; Saito 2009), respectively.

Foraging strategies and feeding habits – a challenge for raptor research

As many raptors rely on a wide prey spectrum, they are often considered to be dietary generalists and opportunistic foragers (e.g., Jaksić & Braker 1983; Thiollay 1994; Sulkava *et al.* 1997). However, raptor diet studies are commonly based on the total niche width of a population and do not consider inter-individual niche variation and specialisation, which has been recognised as a key determinant of a population's niche width (Bolnick *et al.* 2003). Furthermore, studies on raptor foraging strategies considering prey-specific characteristics and local food availability, which allow the assessment of prey choice and dietary shifts on a spatial or seasonal scale, are scarce, particularly in large species such as the white-tailed eagle *Haliaeetus albicilla*. Without such knowledge, it is not possible to determine whether raptor diets reflect opportunistic foraging patterns or the adjustments of selective predators to fluctuations in the availability of prey species with differing profitabilities.

The outcome of dietary investigations heavily depends on the observability of the species of interest (Rosenberg & Cooper 1990). Raptor species are usually difficult to observe as they are afraid of humans and cover relatively large areas during foraging (Austin *et al.* 1996; Väli *et al.* 2004). Consequently, feeding habits are often assessed by indirect means such as the collection of food remains and oral pellets. This type of analysis can be biased because the determination of diet composition is strongly influenced by the digestibility of food items, e.g., hard-bodied versus soft-bodied items (Mersmann *et al.* 1992; Redpath *et al.* 2001). To compensate such a bias, a complementary approach based on the use of stable isotope chemistry can be suitable. This approach relies on the ratios of stable isotopes such as nitrogen ($^{15}N/^{14}N$) or carbon ($^{13}C/^{12}C$) in consumer tissues, which reflect those of their prey in a predictable manner (DeNiro & Epstein 1978, 1981; Frey 2006). Although several studies identified stable isotope analysis as a powerful tool in dietary investigations (e.g., DeNiro & Epstein 1978, 1981; Urton & Hobson 2005; Martínez del Rio *et al.* 2009), it has rarely been attempted in raptors. With respect to lead poisoning, this could be helpful to provide robust estimates of the contribution of potential lead sources to raptor diets.

The white-tailed eagle – a bioindicator for environmental pollutants

Fig. 1.1. *Perching white-tailed eagle; picture by Oliver Krone.*

The white-tailed eagle (Fig. 1.1) belongs to the family Accipitridae, is the largest eagle in Europe and the fourth largest eagle in the world (4.1-6.9 kg, 2.0-2.4 m wing-span; Thiollay 1994). Currently, its distribution extends over the northern Palaearctic and comprises Scandinavia, central and southeast Europe, central and northern Asia, and isolated populations in southwest Greenland, northeast Ireland and western Scotland (Thiollay 1994). Historically, many white-tailed eagle populations were substantially reduced over the entire range, and in numerous western and southern European countries, populations were eradicated as a result of persistent human persecution (Fischer 1982; Helander & Sternberg 2002). In Europe, legal protection started in 1934 and halted the decline of then remaining white-tailed eagle populations but new anthropogenic threats such as land development and chemical pollution emerged (Hauff 1998; Helander & Sternberg 2002). As white-tailed eagles are long-living top-end predators in several habitats, they are considered to be sensitive bioindicators for the accumulation of environmental pollutants (Helander *et al.* 2008). Although conservation measures and the ban of some toxic substances during the last three decades of the 20th century resulted in improved breeding success and increasing populations, especially in northern and central Europe (Thiollay 1994; Hauff 2003), the species still suffers from anthropogenic threats. Today, lead poisoning constitutes a major mortality factor for the global white-tailed eagle population (Kim *et al.* 1999; Krone *et al.* 2006; Helander *et al.* 2009). Species of the genus *Haliaeetus* are in particular susceptible to lead intoxication because the extreme low pH-value (about 1.3) of their gastric acid (Duke *et al.* 1975) and the very long retention period (approximately two days) of lead particles in their stomach (Pattee *et al.* 1981) accelerate the dissolution and subsequent intestinal absorption of lead.

In Germany, a long-term study on the causes of mortality in white-tailed eagles revealed that lead intoxication was the most important cause of death: in 390 white-tailed eagle carcasses collected throughout Germany from 1996 to 2007, 23% were confirmed by necropsy to have died from lead poisoning (Krone *et al.* 2009). Therefore, there is an urgent need to identify the causes of lead poisoning to permit the development of approaches to solve the lead problem.

Framework and purpose of this thesis

In view of the serious threat of lead poisoning to the German white-tailed eagle population, scientists of the Leibniz Institute for Zoo and Wildlife Research Berlin (IZW) conducted a workshop to which all relevant stakeholders, including representatives of several hunting associations, the ammunition industry, ammunition suppliers, foresters, representatives of conservation NGOs, veterinarians and wildlife biologists were invited (Krone & Hofer 2005). The intention of this workshop was to develop an interdisciplinary approach that addresses the concern of each stakeholder group, as their involvement is a valuable step towards resolving human-wildlife conflicts in a sustainable manner (Redpath *et al.* 2004; Menzel & Teng 2010). The participants elaborated a comprehensive catalogue of open questions that needed to be addressed on the causes and potential solutions of lead poisoning in white-tailed eagles in Germany. As a consequence, the cooperative research project “Lead poisoning in white tailed eagles: causes and approaches to solutions” was initiated by the IZW and funded by the Federal Ministry of Education and Research (BMBF, Bundesministerium für Bildung und Forschung).

The present thesis is part of this project and focused on the following two open questions concerning the relationship between the feeding ecology of white-tailed eagles and lead exposure:

(1) What are the sources of lead poisoning in white-tailed eagles?

Although several studies on the diet of white-tailed eagles exist (Oehme 1975; Helander 1983; Sulkava *et al.* 1997), little attention was paid to the relevance of hunter-killed mammal carrion in eagle diet, since the quantification of this food source is very difficult because of its ephemeral nature and unpredictability in space and time. Yet such mammalian carrion is assumed to be the main source of lead fragment and poisoning (Kim *et al.* 1999; Hunt *et al.* 2006).

(2) Why were only small bullet fragments found in gizzards of white-tailed eagles? Do eagles avoid the ingestion of large metal fragments?

Lead bullet fragments isolated from the intestinal tract of lead-poisoned white-tailed eagles were generally small and did not represent the full spectrum of bullet fragment sizes that develop after a lead-based bullet hits an animal (Krone *et al.* 2009). As the handling of food items by white-tailed eagles and other lead-exposed raptors has never been examined in detail before, it is unknown whether this observation reflects selective feeding behaviour in that eagles avoid the ingestion of larger metal fragments. If so, detailed knowledge about the size of avoided metal particles could have substantial implications for the design and application of future bullets.

The purpose of this thesis was to gather comprehensive information on the foraging ecology and feeding habits of white tailed eagles as a model species for large, lead-exposed raptors. More specifically, I addressed this topic from three different points of view:

(1) Within a methodological context, this study presents data from both classical and novel techniques for raptor diet analyses, which enable me to assess the methodological impact on the results of dietary studies and to identify the appropriate approach to study raptor diets.

(2) From a theoretical perspective, the study aims to shed light on the hunting behaviour and foraging strategies of white-tailed eagles in relation to individual variation, fluctuating environments and optimal foraging models.

(3) With the focus on conservation-oriented questions, this thesis provides information necessary for the design of measures to reduce lead poisoning in white-tailed eagles and other raptors with similar feeding habits.

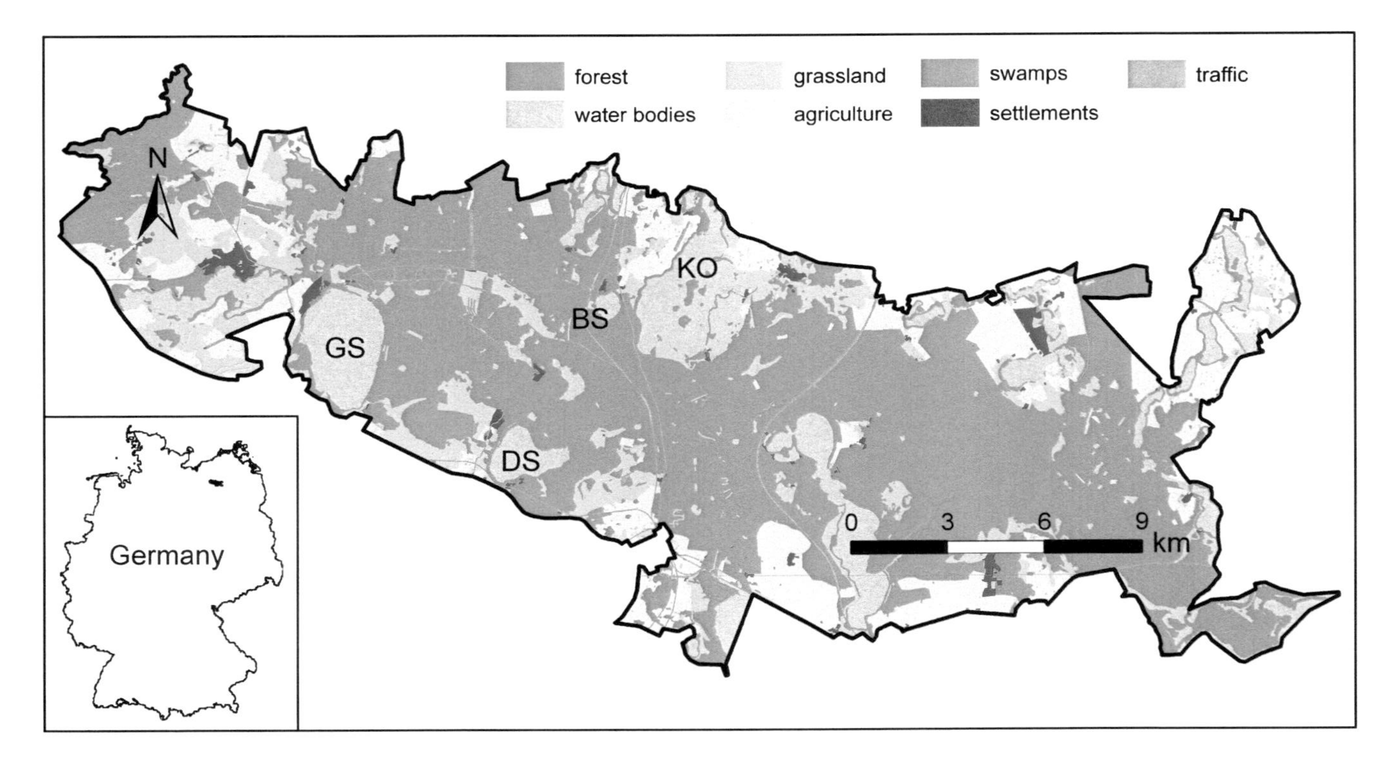

Fig. 1.2. Location and habitat conditions of the nature park Nossentiner/Schwinzer Heide. The following lakes were inhabited by territorial white-tailed eagle pairs on which I focused the data collection: Goldberger See (GS), Damerower See (DS), Bossower See (BS), and Krakower Obersee (KO).

Study site

I conducted my data collection in the nature park *Nossentiner/Schwinzer Heide* situated in northeastern Germany (53°30'-53°40'N, 11°59'-12°35'E; Fig. 1.2). It covers 356 km^2 within the Mecklenburg Lake district which hosts the core of the German white-tailed eagle population. With approximately 15 breeding pairs, white-tailed eagle density in the *Nossentiner/Schwinzer Heide* is one of the highest recorded in Germany (Hauff & Mizera 2006). Since the landscape is characterised by numerous freshwater lakes, a high proportion of pinewoods and mixed coniferous-deciduous forests, extensive livestock pastures and a low human population density, it is regarded as excellent habitat for white-tailed eagles (Fig. 1.2).

Typical for regions within the temperate climate zone, the study site exhibits seasonally changing weather conditions which differ substantially not only between summer and winter but also between spring and autumn. During the study period from January 2007 to December 2008, mean monthly minimum/maximum temperatures ranged from 0.4°C/3.9°C (December) to 12.8°C/22.8°C (July; Fig. 1.3). Rainfall occurred year-round with a maximum in summer and a mean annual precipitation of 664 mm (Fig. 1.3).

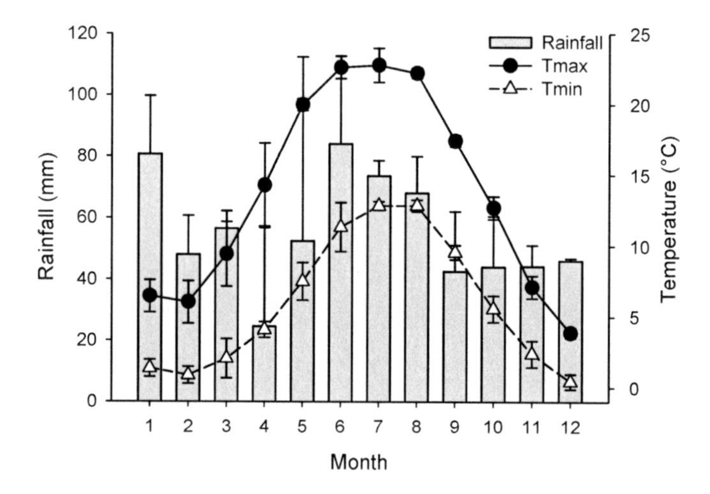

Fig. 1.3. *Precipitation and temperature at the study site during the study period from January 2007 to December 2008. The left and right y-axis values represent mean monthly rainfall and minimum and maximum temperature, respectively, and were measured at the meteorological station "Goldberg" (AWST 3040; located near the Goldberger See within the study site).*

Structure of the thesis

The results of this thesis are presented as four manuscripts in chapters 2 to 5:

Chapter 2 (*"The link between feeding ecology and lead poisoning in white-tailed eagles"*) describes how the feeding ecology of white-tailed eagles results in lead poisoning. Diet was characterised through a combination of classical approaches: the collection of food remains and oral pellets and the analysis of stomach contents. Food availability was based on the monitoring of prey populations and the analysis of hunting bags. I adopted multiple linear regression models to determine the functional (dietary) response of eagles to fluctuating food availabilities and to examine food preferences at prey class level. The analysis of (1) the relationship between seasonal changes in the diet composition and the incidence of lead poisoning in white-tailed eagles, and (2) the stomach contents from lead-poisoned eagles allowed the identification of the main lead sources in this species.

Chapter 3 (*"Sit-and-wait for large prey: foraging strategy and prey choice of white-tailed eagles"*) focuses on the foraging strategy of white-tailed eagles in relation to optimal foraging theory. I observed time allocation of adult territorial white-tailed eagles and foraging behaviour and prey capture success in eagles of all age classes. I employed linear mixed models to identify environmental and individual factors which influence eagle foraging patterns. The results of Chapter 2 concerning food preferences among prey classes were the prerequisite to analyse diet preferences in Chapter 3 by applying a use-availability design. Besides food availability, I considered prey species-specific characteristics to assess the foraging efficiency of white-tailed eagles.

Chapter 4 (*"Specialisation in a generalist predator? Evidence from stable isotope analysis for spatial, seasonal and individual variation in diet composition of white-tailed eagles"*) uses the results of the classical diet analyses in chapters 2 and 3 as background information to assess the use of stable isotopes ($\delta^{13}C$, $\delta^{15}N$) as a promising technique for detecting patterns of dietary variation and quantifying food composition in raptors. I measured stable isotope ratios of tissues from white-tailed eagles from Germany, Finland and Greenland and their potential food species to explore spatial, seasonal, and individual dietary differences on a long-term scale. The use of mass-balance and Bayesian mixing models allowed a robust quantitative assessment of the contribution of potential food sources to the diet of white-tailed eagles.

Chapter 5 (“*Who ingests what and how much? An experimental approach to simulate lead exposure and food processing of white-tailed eagles and other scavengers at shot mammalian carcasses*”) presents an experimental approach to investigate carrion use and the handling of food items of white-tailed eagles and other scavengers in northeastern Germany. I prepared ungulate carcasses with non-toxic iron particles of different sizes, simulating bullet fragments, and offered them as food source to (1) free-ranging scavengers at hunting glades and (2) captive white-tailed eagles. The relative carcass intake by different scavenger species reflected their exposure to lead bullet fragments. The detailed examination of avian feeding behaviour at carcasses containing different-sized metal particles was suitable to test the hypothesis that species such as white-tailed eagles avoid the intake of larger bullet fragments. Since the bullet fragment sizes vary with bullet type, the experimental results could identify bullets that are suitable to prevent metal ingestion and to solve the problem of lead poisoning.

This dissertation concludes with a general discussion in ***chapter 6*** that ties together the findings of all four manuscripts to draw a full picture of the foraging ecology and feeding habits of white-tailed eagles. Here, I considered the methodological implications for dietary studies and evaluated the significance of my thesis for the conservation management of white-tailed eagles with respect to lead poisoning.

References

Austin, G.E., Thomas, C.J., Houston, D.C. & Thomson, D.B.A. 1996. Predicting the spatial distribution of buzzard *Buteo buteo* nesting areas using a Geographical Information System and remote sensing. *Journal of Applied Ecology* 33: 1541-1550.

Beintema, N.H. 2001. *Lead poisoning in waterbirds. International Update Report 2000.* Wetlands International, Wageningen, Netherlands.

Bellrose, F.C. 1959. Lead poisoning as a mortality factor in waterfowl populations. *Illinois Natural History Survey Bulletin* 27: 235-288.

Boere, G.C. 1995. The conservation of migratory birds: the Bonn Convention and the African/Eurasian Waterbird Agreement; a summary of progress and prospects. *Ibis* 137: 214.

Bolnick, D.I., Svanbäck, R., Fordyce, J.A., Yang, L.H., Davis, J.M., Hulsey, C.D. & Forister, M.L. 2003. The ecology of individuals: incidence and implications of individual specialization. *American Naturalist* 161: 1-28.

Côté, I.M. & Sutherland, W.J. 1997. The effectiveness of removing predators to protect bird populations. *Conservation Biology* 11: 395-405.

Darwin, C.R. 1859. *On the origin of species by means of natural selection, or the preservation of favoured races in the struggle for life*. John Murray, London, UK.

DeNiro, M.J. & Epstein, S. 1978. Influence of diet on the distribution of carbon isotopes in animals. *Geochimica et Cosmochimica Acta* 42: 495-506.

DeNiro, M.J. & Epstein, S. 1981. Influence of diet on the distribution of nitrogen isotopes in animals. *Geochimica et Cosmochimica Acta* 45: 341-351.

Duke D.E., Jegers, A.A., Loff, G. & Evanson, O.A. 1975. Gastric digestion in some raptors. *Comparative Biochemistry and Physiology* 50: 649-656.

Emlen, J.M. 1966. The role of time and energy in food preferences. *American Naturalist* 100: 611-617.

Figuerola, J., Mateo, R., Green, A.J., Mondain-Monval, J.-Y., Lefranc, H. & Mentaberre, G. 2005. Grit selection in waterfowl and how it determines exposure to ingested lead shot in Mediterranean wetlands. *Environmental Conservation* 32: 226-234.

Fischer, W. 1982. *Die Seeadler*. Ziemsen Verlag, Wittenberg Lutherstadt, Germany.

Fisher, I.J., Pain, D.J., & Thomas, V.G. 2006. A review of lead poisoning from ammunition sources in terrestrial birds. *Biological Conservation* 131: 421-432.

Fraser, M.D. & Gordon, I.J. 1997. The diet of goats, red deer and South American camelids feeding on three contrasting Scottish upland vegetation communities. *Journal of Applied Ecology* 34: 668-686.

Frey, B. 2006. *Stable isotope ecology*. Springer, New York, USA.

Frey, D.M. 1996. Vulnerability of avian populations to environmental pollutants. *Comments on Toxicology* 5: 401-414.

González, L.M., Margalida, A., Sánchez, R. & Oria, J. 2006. Supplementary feeding as an effective tool for improving breeding success in the Spanish imperial eagle (*Aquila adalberti*). *Biological Conservation* 129: 477-486.

Hauff, P. 1998. Bestandsentwicklung des Seeadlers *Haliaeetus albicilla* in Deutschland seit 1980 mit einem Rückblick auf die vergangenen 100 Jahre. *Vogelwelt* 119: 47-63.

Hauff, P. 2003. Sea-eagles in Germany and their population growth in the 20th century. In: *Sea eagle 2000*, (eds.) Helander, B., Marquiss, M. & Bowerman, B., pp. 71-77. Swedish Society for Nature Conservation/SNF, Stockholm, Sweden.

Hauff, P. & Mizera, T. 2006. Verbreitung und Dichte des Seeadlers *Haliaeetus albicilla* in Deutschland und Polen: eine aktuelle Atlas Karte. *Vogelwelt* 44: 134-136.

Helander, B. 1983. Reproduction of the white-tailed sea eagle *Haliaeetus albicilla* (L.) in Sweden, in relation to food and residue levels of organochlorine and mercury compounds in the eggs. *Ph.D. dissertation*, University of Stockholm, Stockholm, Sweden.

Helander, B. & Sternberg, T. 2002. Action plan for the conservation of white-tailed sea eagles (*Haliaeetus albicilla*). Birdlife International, Strasbourg, France.

Helander, B., Bignert, A. & Asplund, L. 2008. Using raptors as environmental sentinels: monitoring the white-tailed sea eagle *Haliaeetus albicilla* in Sweden. *Ambio* 37: 425-431.

Helander, B., Axelsson, J., Borg, H., Holm, K. & Bignert, A. 2009. Ingestion of lead from ammunition and lead concentrations in white-tailed sea eagles (*Haliaeetus albicilla*) in Sweden. *Science of the Total Environment* 407: 5555-5563.

Hunt, W.G., Burnham, W., Parish, C.N., Burnham, B., Mutch, B. & Oaks, J.L. 2006. Bullet fragments in deer remains: implications for lead exposure in scavengers. *Wildlife Society Bulletin* 34: 167-170.

Jaksić, F.M. & Braker, H.E. 1983. Food-niche relationships and guild structure of diurnal birds of prey: competition versus opportunism. *Canadian Journal of Zoology* 61: 2230-2241.

Kenntner N., Tataruch, F. & Krone, O. 2001. Heavy metals in soft tissue of white-tailed eagles found dead or moribund in Germany and Austria from 1993 to 2000. *Environmental Toxicology and Chemistry* 20: 1831-1837.

Kim, E.Y., Goto, R., Iwata, H., Masuda, Y., Tanabe, S. & Fujita, S. 1999. Preliminary survey of lead poisoning of Steller's sea eagle (*Haliaeetus pelagicus*) and white-tailed sea eagle (*Haliaeetus albicilla*) in Hokkaido, Japan. *Environmental Toxicology and Chemistry* 18: 448-451.

Korine, C., Izhaki, I. & Arad, Z. 1999. Is the Egyptian fruit-bat *Rousettus aegyptiacus* a pest in Israel? An analysis of the bat's diet and implications for its conservation. *Biological Conservation* 88: 301-306.

Krebs, J.R. 1978. Optimal foraging: decision rules for predators. In: *Behavioral ecology: an evolutionary approach*, (eds.) Krebs, J.R. & Davies, N.B., pp. 23-63. Blackwell Scientific Publications, Oxford, UK.

Krone, O. & Hofer, H. (eds.). 2005. *Bleihaltige Geschosse in der Jagd – Todesursache von Seeadlern*? Leibniz Institute for Zoo and Wildlife Research, Berlin, Germany.

Krone, O., Stjernberg, T., Kenntner, N., Tataruch, F., Koivusaari, J., & Nuuja, I. 2006. Mortality, helminth burden and contaminant residues in white-tailed sea eagles from Finland. *Ambio* 35: 98-104.

Krone, O., Kenntner, N. & Tataruch, F. 2009. Gefährdungsursachen des Seeadlers (*Haliaeetus albicilla* L. 1758). *Denisia* 27: 139-146.

Locke, L.N. & Thomas, N.J. 1996. Lead poisoning of waterfowl and raptors. In: *Noninfectious diseases of wildlife*, (eds.) Fairbrother, A., Locke, L.N. & Hoff, G.L., pp. 108-117. Iowa State University Press, Iowa, USA.

López-Bao, J.V., Alejandro Rodrígueza, A. & Palomaresa, F. 2008. Behavioural response of a trophic specialist, the Iberian lynx, to supplementary food: patterns of food use and implications for conservation. *Biological Conservation* 141: 1857-1867.

Lumeij, J.T. 1985. Clinicopathologic aspects of lead poisoning in birds: a review. *Veterinary Quarterly* 7: 133-138.

Martina, P.A., Campbell, D., Hughes, K. & McDaniel, T. 2008. Lead in the tissues of terrestrial raptors in southern Ontario, Canada, 1995-2001. *Science of the Total Environment* 391: 96-103.

Martínez del Rio, C., Sabat, P., Anderson-Sprecher, R. & Gonzalez, S.P. 2009. Dietary and isotopic specialization: the isotopic niche of three *Cinclodes* ovenbirds. *Oecologia* 161: 149-159.

Mateo, R., Green, A.J., Jeske, C.W., Urios, V. & Gerique, C. 2001. Lead poisoning in the globally threatened marbled teal and white-headed duck in Spain. *Environmental Toxicology and Chemistry* 20: 2860-2868.

Mateo, R., Green, A.J., Lefranc, H., Baos, R. & Figuerola, J. 2007. Lead poisoning in wild birds from southern Spain: a comparative study of wetland areas and species affected, and trends over time. *Ecotoxicology and Environmental Safety* 66: 119-126.

McArthur, R.H. & Pianka, E. 1966. On optimal use of a patchy environment. *American Naturalist* 100: 603-609.

Menzel, S. & Teng, J. 2009. Ecosystem services as a stakeholder-driven concept for conservation science. *Conservation Biology* 24: 907-909.

Meretsky, V.J., Snyder, N.F.R., Beissinger, S.R., Clendenen, D.A. & Wiley, J.W. 2000. Demography of the California condor: implications for reestablishment. *Conservation Biology* 14: 957-967.

Mersmann, T.J., Buehler, D.A., Fraser, J.D. & Seegar, J.K.D. 1992. Assessing bias in studies of bald eagle food habits. *Journal of Wildlife Management* 56: 73-78.

Miller, M.J.R., Wayland, M.E. & Bortolotti, G.R. 2002. Lead exposure and poisoning in diurnal raptors: a global perspective. In: *Raptors in the New Millenium*, (eds.) Yosef, R.,

Miller, M.L. & Pepler, D., pp. 224-245. International Birding and Research Center, Eilat, Israel.

Oehme, G. 1975. Ernährungsökologie des Seeadlers, *Haliaeetus albicilla* (L.), unter besonderer Berücksichtigung der Population in den drei Nordbezirken der Deutschen Demokratischen Republik. *Doctoral dissertation*, Universität Greifswald, Germany.

Pain, D.J. 1990. Lead shot ingestion by waterbirds in the Camargue, France: an investigation of levels and interspecific differences. *Environmental Pollution* 66: 273-285.

Pain, D.J., Sears, J. & Newton, I. 1995. Lead concentrations in birds of prey in Britain. *Environmental Pollution* 87: 173-180.

Pattee, O.H., Wiemeyer, S.N., Mulhern, B.M., Sileo, L. & Carpenter, J.W. 1981. Experimental lead-shot poisoning in bald eagles. *Journal of Wildlife Management* 45: 806-810.

Pyke, G.H., Pulliam, H.R. & Charnov, E.L. 1977. Optimal foraging: a selective review of theory and tests. *The Quarterly Review of Biology* 52: 137-154.

Rattner, B.A. 2009. History of wildlife toxicology. *Ecotoxicology* 18: 773-783.

Redpath, S.M., Clarke, R., Madders, M. & Thirgood, S.J. 2001. Assessing raptor diet: comparing pellets, prey remains, and observational data at hen harrier nests. *Condor* 103: 184-188.

Redpath, S.M., Arroyo, B.E., Leckie, F.M., Bacon, P., Bayfield, N., Gutiérrez, R.J. & Thirgood, S.J. 2004. Using decision modeling with stakeholders to reduce human-wildlife conflict: a raptor-grouse case study. *Conservation Biology* 18: 350-359.

Rosenberg, K.V. & Cooper, R.J. 1990. Approaches to avian diet analysis. *Studies in Avian Biology* 13: 80-90.

Saito, K. 2009. Lead poisoning of Steller's sea eagle (*Haliaeetus pelagicus*) and white-tailed eagle (*Haliaeetus albicilla*) caused by the ingestion of lead bullets and slugs, in Hokkaido, Japan. In: *Ingestion of lead from spent ammunition: implications for wildlife and humans*, (eds.) Watson, R.T., Fuller, M., Pokras, M. & Hunt, G., pp. 302-309. The Peregrine Fund, Idaho, USA.

Sanderson, G.C. & Bellrose, F.C. 1986. A review of the problem of lead poisoning in waterfowl. *Illinois Natural History Survey Special Publication* 4: 1-34.

Scheuhammer, A.M. 1987. The chronic toxicity of aluminium, cadmium, mercury, and lead in birds: a review. *Environmental Pollution* 46: 263-295.

Scheuhammer, A.M. & Norris, S.L. 1996. The ecotoxicology of lead shot and lead fishing weights. *Ecotoxicology* 5: 279-295.

Scheuhammer, A.M. & Templeton, D.M. 1998. Use of stable isotope ratios to distinguish sources of lead exposure in wild birds. *Ecotoxicology* 7: 37-42.

Scheuhammer, A.M., Meyer, M.W., Sandheinrich, M.B. & Murray, M.W. 2007. Effects of environmental methylmercury on the health of wild birds, mammals, and fish. *Ambio* 36: 12-19.

Schoener, T.W. 1971. Theory of feeding strategies. *Annual Review of Ecology and Systematics* 2: 369-404.

Sibly, R.M., Newton, I. & Walker, C.H. 2000. Effects of dieldrin on population growth rates of sparrowhawks 1963-1986. *Journal of Applied Ecology* 37: 540-546.

Sieg, R., Sullivan, K.A. & Parish, C.N. 2009. Voluntary lead reduction efforts within the northern Arizona range of the California condor. In: *Ingestion of lead from spent ammunition: implications for wildlife and humans*, (eds.) Watson, R.T., Fuller, M., Pokras, M. & Hunt, G., pp. 341-349. The Peregrine Fund, Idaho, USA.

Sinclair, A.R.E., Freyxell, J.M. & Caughley, G. 2006. *Wildlife ecology, conservation, and management*. 2nd edition. Blackwell Science, Oxford, UK.

Stephens, D.W. & Krebs, J.R. 1986. *Foraging theory*. Princeton University Press, Princeton, USA.

Sulkava, S., Tornberg, R. & Koivusaari, J. 1997. Diet of the white-tailed eagle *Haliaeetus albicilla* in Finland. *Ornis Fennica* 74: 65-78.

Thiollay, J.M. 1994. Family Accipitridae (hawks and eagles). In: *Handbook of the birds of the world: New World vultures to guineafowl*. Volume 2, (eds.) del Hoyo, J., Elliot, A. & Sargatal, J., pp. 52-205. Lynx Edicions, Barcelona, Spain.

Thomas, V.G. & Owen, M. 1996. Preventing lead toxicosis of European waterfowl by regulatory and nonregulatory means. *Environmental Conservation* 23: 358-364.

Thomas, V.G. & Guitart, R. 2005. Role of international conventions in promoting avian conservation through reduced lead toxicosis: progression towards a non-toxic agenda. *Bird Conservation International* 15: 147-160.

Urton, E.J.M. & Hobson, K.A. 2005. Intrapopulation variation in grey wolf isotope ($\delta^{15}N$ and $\delta^{13}C$) profiles: implications for the ecology of individuals. *Oecologia* 145: 317-326.

Väli, Ü., Treinys, R. & Lõhmus, A. 2004. Geographical variation in macrohabitat use and preferences of the lesser spotted eagle *Aquila pomarina*. *Ibis* 146: 661-671.

Williams, T.M., Estes, J.A., Doak, D.F. & Springer, A.M. 2004. Killer appetites: assessing the role of predators in ecological communities. *Ecology* 85: 3373-3384.

CHAPTER 2

The link between feeding ecology and lead poisoning in white-tailed eagles

Abstract

Lead poisoning affects numerous threatened raptors and is the major cause of death in white-tailed eagles *Haliaeetus albicilla*. The primary reason for intoxication is assumed to be lead fragments ingested while feeding on game mammals killed by lead-based bullets. However, empirical evidence on the relevance of carrion in raptor diets remains scarce. We therefore investigated the link between raptor feeding ecology and lead poisoning, with white-tailed eagles as a model species for scavenging birds. Data on seasonal diet composition of seven territorial white-tailed eagle pairs were collected during a two-year field period in northeastern Germany. The relationship between seasonal changes in diet and food availability was examined in five eagle pairs by using multiple linear regression models. We also analysed stomach contents of 126 eagles found dead all over Germany. Our results revealed that fish were the primary prey for eagles, and waterfowl and carcasses of game mammals were important alternative diet components. Eagles used individual foraging tactics, adjusted to local food supply, in such a way that they maximised profitability. They showed a type II functional response to fish availability. When fish availability sharply declined, eagles switched to waterfowl and carrion. Correspondingly, diet breadth of study pairs increased with decreasing fish availability. The consumption of game mammal carrion increased over autumn and winter and was positively correlated with a concomitant seasonal increase in the incidence of lead poisoning in eagles throughout Germany. The stomachs of lead-poisoned eagles predominantly contained ungulate remains. These results indicate that carcasses of game mammals were the major sources of lead fragments. The link between raptor feeding ecology and lead poisoning is the specific functional response of raptors to changing food availability or poor habitat quality, leading to scavenging on lead-contaminated carrion. As alternative food source, carrion constitutes a considerable threat to white-tailed eagles and other birds with similar feeding habits as long as it contains lead bullet fragments. Conservation management of scavenging birds would be substantially improved if carrion was free of lead bullet fragments. One method to achieve this is the widespread introduction of lead-free ammunition.

Keywords: diet, food availability, functional response, *Haliaeetus albicilla*, lead poisoning, lead sources, niche breadth, raptor conservation, seasonal variation

Introduction

Oral lead intoxication constitutes an anthropogenic, potentially lethal threat to wildlife, with negative effects on general health, reproduction and behaviour (Fisher *et al.* 2006). One of the main groups affected by lead poisoning is raptors (Martina *et al.* 2008). High tissue lead levels have been documented in 35 species from 18 countries (Miller *et al.* 2002). In white-tailed eagles *Haliaeetus albicilla*, lead poisoning is a significant mortality factor in several countries (Kim *et al.* 1999; Helander *et al.* 2009) and the most important cause of death in Germany (Krone *et al.* 2009*a*). The geographic range of this largest European eagle extends over the northern Palaearctic. Intensive persecution led to the extinction of white-tailed eagle populations in many parts of Europe until the beginning of the 20th century (Fischer 1982). In spite of subsequent conservation efforts, population growth was repeatedly prevented by toxins, pollutants and human disturbance. Presently, the population in northern and central Europe is increasing, from a low of about 15 breeding pairs in 1913 (Glutz von Blotzheim *et al.* 1971) to approximately 570 breeding pairs in Germany alone in 2007 (Hauff 2008). The population is still in the process of recovery (Sulawa *et al.* 2010) and the German population spearheads the resettlement of western Europe because the western distribution limit is in northeastern Germany (Hauff 2003).

Knowledge about the diet composition of birds of prey is essential for their conservation management and offers information on the proximate sources of toxins such as metallic lead particles (Kenntner *et al.* 2001; Sinclair *et al.* 2006). The main lead sources are assumed to be tissues of shot game animals interspersed with fragments of lead shot or lead-based bullets (Pain & Amiard-Triquet 1993; Thomas & Owen 1996; Scheuhammer & Templeton 1998; Martina *et al.* 2008). It has been suggested that contaminated carcasses of mammals and gut piles provide a higher exposure to lead intoxication in white-tailed eagles and other avian scavengers than waterfowl do (Kim *et al.* 1999; Meretsky *et al.* 2000; Hunt *et al.* 2006). Dietary investigations showed that white-tailed eagles prey on fish, birds and mammals and feed on carrion (Helander 1983; Sulkava *et al.* 1997; Struwe-Juhl 2003). However, the relative contribution of mammals and carrion to the diet of white-tailed eagles remains unclear, with reported results of between 3 % (Struwe-Juhl 2003) and 34 % (Helander 1983).

In previous studies, the proportion of fish in the diet of white-tailed eagles increased in summer and that of waterfowl in winter (Oehme 1975; Helander 1983; Struwe-Juhl 2003). To what extent seasonal fluctuations in the diet are linked to food availability, reflect the main food sources in a season and influence the uptake of lead has not been examined. This is important because the number of lead-poisoned eagles significantly changed with season and showed a peak in winter during the main hunting season (Krone *et al.* 2009*a*). Functional (dietary) responses to changes in prey abundance have been documented for several raptor species (Sonerud 1992). Optimal foraging models are generally based on the hypothesis that an animal's pattern of food choice will be such that the net energy gained per unit time spent foraging is maximised (Schoener 1971). This leads to the prediction that diet breadth increases with a decrease in the availability of preferred prey (Pyke *et al.* 1977). We therefore expected that – on the basis of published results – when the availability of fish and waterfowl is scarce, white-tailed eagles should respond by feeding on alternative food types such as carrion.

The aim of our study was to investigate how the feeding ecology of individual raptors results in lead poisoning, with white-tailed eagles as a model species for scavenging avian species. We therefore determined (1) the importance of mammals/carrion in the diet of white-tailed eagles across seasons, (2) the functional response of eagles to changing food availability, (3) the extent of dietary shifts realised by eagles, and (4) the major sources of lead fragments.

Material and methods

Study site and study animals

The study was conducted in the nature park *Nossentiner/Schwinzer Heide*, located in the lake district of Mecklenburg-Western Pomerania in northeastern Germany (53°30'-53°40'N, 11°59'-12°35'E). This bird sanctuary contains, with 15 breeding pairs on 356 km^2, a white-tailed eagle population with one of the highest densities recorded in Germany. The area is characterised by 60 freshwater lakes covering 14 % of the landscape, a high proportion of secondary forests (60 %) and agriculture (21 %) and a low proportion of settlement and traffic (5 %). Study animals were seven territorial white-tailed eagle pairs (SP1 to SP7, Table 2.1). In the case of pairs SP1, SP3, SP4, SP6 and SP7, one individual was fitted with a GPS-transmitter to study space and habitat use (for further details see Scholz 2010). The home

Table 2.1. *The study pairs and the lakes in their home range most frequently used for foraging activities.*

Study pairs (SP)	Main water body for eagle foraging*	Area (ha)	Maximum depth (m)	Mean depth (m)	Trophic state
SP1 (KO East) SP2 (KO West) SP3 (KO North)	*Krakower Obersee* (KO)	799	28.3	7.5	eutrophic (e2)
SP4	*Damerower See* (DS)	285	7.0	2.0	polytrophic
SP5	*Goldberger See* (GS)	770	4.1	2.1	polytrophic
SP6	*Bossower See* (BS)	54	8.8	3.4	eutrophic (e2)
SP7	*Schwarzer See* (SS)**	2	10.9	6.0	weak eutrophic (e1)

*The lake where the majority of total prey items were found, up to 100 m away from the littoral zone.

**The SS was not a main water body but merely a minor water body for foraging; still most prey items (21 %) of SP7 were found at this lake.

ranges of all study pairs except for pair SP7 contained at least one bigger lake (≥ 54 ha), on which the eagles focused their foraging. The home range of pair SP7 included no main water body and only two small lakes (≤ 2 ha). Pair SP7 regularly went on foraging excursions to other lakes outside their home range.

In addition, from a long-term study on the causes of mortality in white-tailed eagles (Krone *et al.* 2003; Krone *et al.* 2009*a*), stomach contents (SC) of 126 individuals that died throughout Germany were examined for food remains.

Data collection

Data were collected from January to December 2007 on pairs SP1 to SP4 and from January to December 2008 on pairs SP5 to SP7. We collected food remains and pellets of all study pairs every month. We systematically searched the home ranges of the study pairs, aided by GPS-positions and feeding sites and roosts known from direct observations. The GPS-transmitters were adjusted to receive at least one position per day during daylight on a shifting schedule with a delay of one hour per day. Except for the breeding season, we searched the vicinity of eyries of all study pairs for food remains and pellets. Nestlings of study pairs were ringed. We only included fresh food items or pellets from eyries in our sample. SP1 to SP5 were each watched in their foraging areas on two days per month. From dawn to dusk, behaviour sampling combined with continuous recording was used to record search effort (i.e., time spent foraging for prey until capture), prey captures and handling time (i.e., time spent from prey capture until consumption; Martin & Bateson 1993). In order to record handling time of white-tailed eagles for carrion, we provided ungulate carcasses and gut piles at six different eagle home ranges in the study area and observed them via video surveillance (M. Nadjafzadeh *et al.* unpublished data).

The 126 stomach contents were from dead eagles which had been collected over the period 2000 to 2008 and submitted for necropsy to the Leibniz Institute for Zoo and Wildlife Research Berlin (IZW).

Food availability for pairs SP1 to SP5 was measured once a month by random sampling. We defined relative food availability as the abundance of the main food types in areas accessible to eagles. The relative abundance of fish in water depths accessible to eagles (ca. 0-1 m; Fischer 1982) in the main water bodies where we observed eagles foraging, the *Krakower Obersee* (KO), *Damerower See* (DS) and *Goldberger See* (GS), was assessed in collaboration with local fishermen. We counted the contents of their fish traps (12 to 15 in each lake) and classified their catch qualitatively by species and quantitatively by number. Mesh size of used fish traps was 14 to 24 mm; only fish $\geq$ 10 cm were captured. From November to February, fishing was interrupted because most fish staid in the lower strata of lakes and were mostly inactive owing to cold water temperatures (García-Berthou 2001; Hanson *et al.* 2008). Water temperatures of KO, DS and GS were measured at the surface and at depths of 1 m and 2 m once a month with a Horiba U-10 Water Quality Checker. During the study period, none of the lakes froze. We estimated fish availability during winter months by using a linear regression on log-transformed data of monthly water temperatures and fish counts in the respective lakes (KO: $r^2 = 0.81$, $P = 0.006$, $N = 7$; DS: $r^2 = 0.81$, $P = 0.006$, $N = 7$; GS: $r^2 = 0.88$, $P = 0.001$, $N = 7$). The relative abundance of waterfowl (i.e., species that belonged to the order Anseriformes and to the families Podicipedidae, Phalacrocoracidae and Rallidae) on the part of lake surfaces accessible to eagles (Rudebeck 1950; Oehme 1975) and used for eagle foraging was recorded by direct observations (Bibby 1995): we counted all waterfowl visible on KO, DS, and GS by using two to six locations as simultaneous recording stations, respectively. The relative abundance of mammal carcasses and gut piles left by hunters in the field and accessible to eagles (Hunt *et al.* 2006) was estimated from hunting bags of the home ranges of all study pairs. These data were provided by the resident forestry districts and the hunters in charge of local hunting blocks.

Data analysis

For taxonomic identification of collected food items, pellet and stomach contents, we prepared a reference collection of bones, scales, feathers and hairs. We also used reference material of the Berlin Museum of Natural History and published identification keys (Teerink 1991). Hair examination was aided by a scanning electron microscope. Several pellets found at a single roost at the same time were treated as a single pellet because remains of a given food item may appear in more than one pellet (Mersmann *et al.* 1992). The minimum number of food items was recorded.

Diet composition was analysed on a yearly and monthly basis and summarised over seasons. Seasons were defined as winter (December-February), spring (March-May), summer (June-August), autumn (September-November), winter half-year (September-February) and summer half-year (March-August). The relative contribution of each food type to the diet of each study pair was estimated as $p_j = F_j / n_t$, where F_j is the number of occurrences of food item j recorded from food remains and pellets, and n_t the total number of all food items collected throughout the year. The relative contribution of each food type to the monthly diet of each pair was $p_{ij} = F_{ij} / n_i$, where F_{ij} is the number of occurrences of food item j in the month i, and n_i the sum of food items per month ($n_i = \sum_{j=1}^{3} F_{ij}$). For the analysis of seasonal diet composition, we pooled all food items over the respective months. We also calculated the relative monthly consumption of each food type over the entire year as $q_{ij} = F_{ij} / F_j$. For individuals whose stomach contents were analysed, we recorded the number of occurrences of each food item per stomach. Diet composition of eagles with different causes of death were calculated as $p_{dj} = F_{dj} / n_d$, where F_{dj} is the number of occurrences of food item j recorded from stomachs of eagles with the cause of death d, and n_d the total number of food items recorded from stomachs of eagles with the cause of death d.

Relative monthly availability of each food type over the entire year was estimated for the home ranges of pairs SP1 to SP5 as $a_{ij} = V_{ij} / V_j$, where V_{ij} is the monthly sample count of food type j, and V_j the total sample count of food type j summed over the entire year.

We calculated diet breadth on a yearly (B) and monthly (B_m) basis using Levins's formula as $B = 1 / \Sigma p_j^2$ and $B_m = 1 / \Sigma p_{ij}^2$ (Krebs 1999). Measures were based on the frequencies of individual species identified at least to genus level in the diet, as recommended by Greene and Jaksić (1983).

For the purpose of data analysis we defined as game mammals species that were included in the hunting bags and belonged to the families Cervidae, Bovidae, Suidae, Canidae, Mustelidae, Procyonidae and Leporidae.

The energy intake rate of white-tailed eagles was calculated as a function of search effort as $R = E_f / (x + T_h)$, where E_f is the net amount of energy gained in the total time spent searching for and handling food, x the minimum and maximum observed search effort and T_h the mean observed handling time (Stephens & Krebs 1986). The net amount of energy was estimated based on the average mass of consumed prey species/carrion. We used 1.3 kcal g^{-1} for fish and 1.9 kcal g^{-1} for waterfowl and ungulate carcasses/gut piles (Helander 1983). Since it was not possible to directly observe the search effort of eagles for carrion, we used the minimum and maximum search effort observed for fish and waterfowl for the plot equation. We considered this as conservative approach, because the main water bodies used for foraging for fish and waterfowl by the study pairs were approximately seven times closer to their eyries than the locations where we documented the consumption of carrion. Furthermore, excursions of study pairs out of their home ranges were often directed to forests, grassland and agriculture where carrion could potentially be found and the maximum distance tracked eagles were observed to cover during such movements was 39 km (Scholz 2010).

All statistical tests were two-tailed and carried out with SPSS 16.0 (SPSS Inc., Chicago, IL, USA). Significance threshold was set at $P < 0.05$. Descriptive statistics are presented as mean ± SD and regression coefficients as mean ± SE. In all cases where we analysed frequencies based on count data we employed the log-likelihood ratio test. Seasonal changes in diet composition and food availability of study pairs were tested with the Friedman test. For the purpose of comparing diet breadth of study pairs between winter half-year and summer half-year, we applied the Wilcoxon signed-ranks test. We examined the relationship between the monthly variation in the consumption of game mammals or waterfowl by eagles (i.e., q_{ij}) and the monthly variation in the number of lead-poisoned eagles by calculating Spearman's rank correlation coefficient ρ. We used ρ to test the relationship between monthly variation in diet breadth (B_m) and (1) diet composition (p_{ij}) and (2) food availability (a_{ij}). We used multiple linear regression models to estimate the relationship between diet and food availability: The mean monthly proportions of fish, waterfowl, and game mammals in the diet (p_{ij}) of SP1 to SP5 represented the dependent variable and the mean monthly availability of fish, waterfowl and carrion in the home ranges (a_{ij}) of SP1 to SP5 were predictor variables. We log-transformed the values of the dependent and the predictor variables to ensure there were no significant deviations of the residuals from normality as judged by the Lilliefors test. All possible subsets of predictor variables were assessed and ranked using Akaike's Information Criterion corrected for small sample sizes AIC_c to identify the most parsimonious model with the best fit (Burnham & Anderson 2002). We used the proportion of variance explained by the model (adjusted R^2) as a general measure of goodness-of-fit.

Results

Diet composition of study pairs

The diet was composed of fish (predominantly Cyprinidae), birds (primarily waterfowl) and mammals (mostly game species, Table 2.2). The study pairs significantly differed in the relative contribution of the three major food types to their diet (log-likelihood ratio test, $G = 70.0$, $df = 12$, $P < 0.0001$, $N = 705$). This was mostly due to SP7. Once this pair was removed, pairs SP1 to SP6 showed no significant differences any more ($G = 17.7$, $df = 10$, $P = 0.063$, $N = 635$). Fish accounted for 65.1 ± 7.8 % and 27.1 %, birds for 21.9 ± 6.4 % and 24.3 %, and mammals for 13.0 ± 2.1 % and 48.6 % in the diet of SP1 to SP6 and SP7, respectively. The relative contribution of different food types across all study pairs varied significantly between months (Friedman test, fish: $\chi^2 = 74.2$, $df = 11$, $P < 0.0001$, $N = 7$; birds: $\chi^2 = 57.2$, $df = 11$, $P < 0.0001$, $N = 7$; mammals: $\chi^2 = 66.9$, $df = 11$, $P < 0.0001$, $N = 7$). Fish were the main prey during spring, with a peak of 92.1 ± 8.1 % in April, birds were most frequently consumed in summer, with a peak of 35.7 ± 17 % in August, and mammals were important in autumn and winter, with a peak of 44.1 ± 14.4 % in December (Fig. 2.1).

Diet breadth differed between study pairs. SP4 and SP5 had the most specialised diet, diets of SP1 to SP3 and SP6 were intermediate in breadth and SP7 had the most diverse diet (Table 2.2). Diet breadth of all pairs increased significantly in winter half-year (Wilcoxon signed-ranks test, $P_{exact} = 0.016$, $N = 7$). Mean monthly diet breadth of SP1 to SP7 increased as the monthly proportion of fish declined ($\rho = -0.93$, $P < 0.0001$, $N = 12$) and increased with the monthly proportion of birds ($\rho = 0.65$, $P = 0.022$, $N = 12$) and mammals ($\rho = 0.85$, $P = 0.001$, $N = 12$) in the diet.

Table 2.2. *Percentages of the frequency of occurrence of different food types in the diet of seven study pairs of white-tailed eagles in northeastern Germany and in the stomach contents of 126 dead white-tailed eagles throughout Germany.*

Measurement	Study pairs							Eagle stomach contents
	SP1	SP2	SP3	SP4	SP5	SP6	SP7	
A) Food type (% of items)								
Fish	59.4	58.7	62.5	75.9	74.0	59.4	27.1	32.8
Cyprinidae	52.8	53.2	51.8	75.9	70.0	56.3	22.9	28.2
Percidae	0.9	0.9	0.9	0	4.0	1.0	4.3	1.7
Esox lucius	5.7	4.6	9.8	0	0	2.1	0	1.7
Salmo spec.	0	0	0	0	0	0	0	0.6
Unidentified fish	0	0	0	0	0	0	0	0.6
Birds	25.5	26.6	25.9	14.3	13.0	27.1	24.3	25.3
Podicipediformes*	3.8	0.9	0.9	5.4	2.0	2.1	0	2.9
*Phalacrocorax carbo**	2.8	1.8	5.4	0	0	1.0	1.4	0
Ardea cinerea	0	0	0	0	0	1.0	0	0
Anseriformes*	8.5	7.3	9.8	7.1	8.0	5.2	4.3	8.0
Rallidae*	8.5	13.8	8.0	0	1.0	15.6	10.0	5.7
Charadriiformes	1.9	1.8	1.8	1.8	1.0	1.0	1.4	1.1
Passeri	0	1.0	0	0	0	0	1.4	0
Columba spec.	0	0	0	0	1.0	1.0	2.9	1.1
Dendrocopos major	0	0	0	0	0	0	1.4	0
Unidentified birds	0	0	0	0	0	0	1.4	6.3
Mammals	15.1	14.7	11.6	9.8	13.0	13.5	48.6	42.0
Cervidae**	6.6	3.7	4.5	4.5	4.0	6.3	14.3	13.2
Bovidae**	0	0.9	0	0	0	0	1.4	0.6
Suidae**	4.7	6.4	4.5	3.6	5.0	5.2	10.0	17.2
Canidae**	0.9	0.9	0.9	1.8	3.0	2.1	7.1	2.9
Mustelidae**	0	0	0	0	0	0	2.9	1.1
*Procyon lotor***	0	0	0	0	0	0	1.4	0
Felis catus	0	0	0	0	1.0	0	0	0
*Lepus europeus***	0.9	0.9	1.8	0	0	0	2.9	3.4
Sciurus vulgaris	0	0	0	0	0	0	2.9	0
Rodentia	1.9	1.8	0	0	0	0	5.7	2.9
Unidentified mammals	0	0	0	0	0	0	0	0.6
B) Diet Breadth (B)	6.6	5.7	5.6	2.1	2.6	6.9	14.4	10.4
C) No. of items (N)	106	109	112	112	100	96	70	174

*included in definition of "waterfowl"; **included in definition of "game mammals"

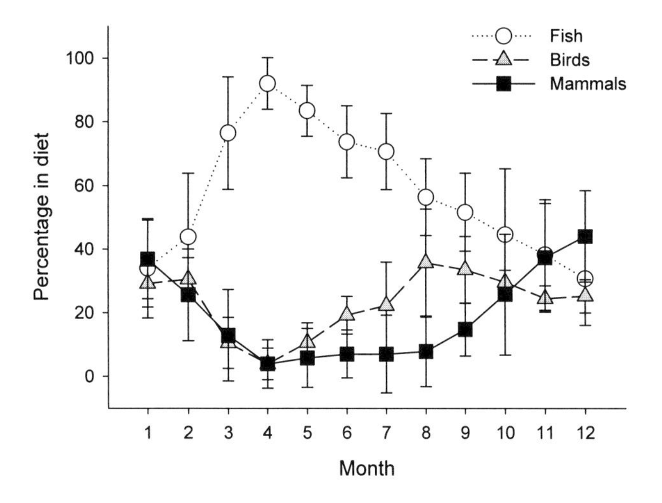

Fig. 2.1. *Seasonal diet of seven study pairs of white-tailed eagles in northeastern Germany as documented by prey remains and pellet contents (N = 705). Error bars denote ±SD.*

Diet composition of eagles with different causes of death

The diet of 126 eagles found dead was, based on stomach contents (SC), diverse and resulted in a wide diet breadth (Table 2.2). The relative contribution of the three major food types to the diet varied significantly with the cause of death ($G = 49.9$, $df = 8$, $P < 0.0001$, $N = 149$). Eagles that died from natural causes or traumas had most frequently consumed fish (Fig. 2.2). Their diet did not differ significantly from the diet of SP1 to SP6 ($G = 22.2$, $df = 14$, $P = 0.083$, $N_{SP1-6} = 635$, $N_{SC} = 79$) but significantly from the diet of SP7 ($G = 15.6$, $df = 4$, $P < 0.005$, $N_{SP7} = 70$, $N_{SC} = 79$). The stomachs of train accident victims and poisoned eagles most frequently contained mammal remains. Pb-poisoned eagles had, with 81 %, a significantly higher proportion of mammals in their diet than expected from the diet composition of eagles with natural causes of death ($G = 14.4$, $df = 2$, $P < 0.001$, $N = 42$); all mammal remains were game, mainly ungulates. The relative contribution of the three major food types in this sample significantly changed with season ($G = 25.5$, $P < 0.001$, $N = 174$), with an increased contribution of mammals in autumn and winter.

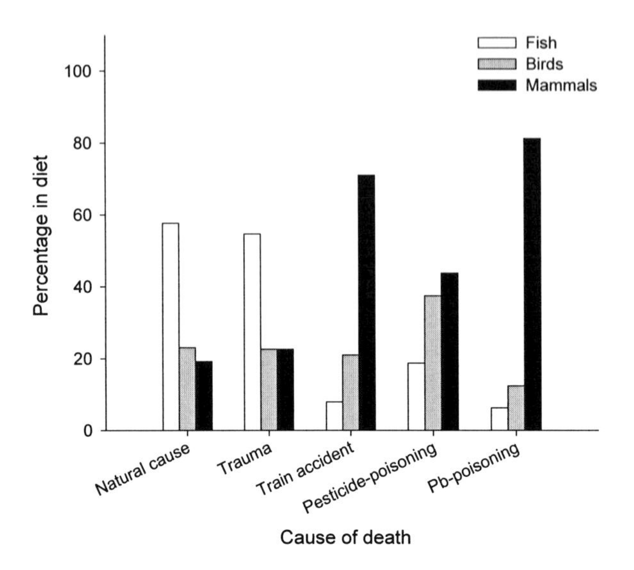

Fig. 2.2. *Diet of white-tailed eagles based on food remains in stomachs of 109 eagles with known causes of death (N = 149).*

Mammals in the diet

Eagles mainly fed on game mammals which comprised 92.8 ± 7.3 % of mammal remains of study pairs (MR_{SP}) and 93.1 % of mammal remains in stomach contents of dead individuals (MR_{SC}). They consumed significantly more large mammals with a body mass above than below 10 kg (study pairs: $P_{exact} = 0.016$, N = 7; stomach contents: G = 18.0, df = 1, P < 0.0001, $N_{<10kg} = 54$, $N_{>10kg} = 18$): wild boar constituted 35.3 ± 7.5 % of MR_{SP} and 41.7 % of MR_{SC} and cervids 37 ± 8.6 % of MR_{SP} and 31.9 % of MR_{SC}. Whenever we found remains of young wild boars and fawns, their bones exhibited bullet wounds or fractures. Amongst game mammals with less than 10 kg body mass, canids (13.1 ± 6.5 % of MR_{SP}, 6.9 % of MR_{SC}) and hares (4.8 ± 5.6 % of MR_{SP}, 8.3 % of MR_{SC}) were regularly consumed.

Functional response to seasonal changes in food availability

Relative food availability for pairs SP1 to SP5 varied significantly between months (fish: χ^2 = 54.7, df = 11, P < 0.0001, N = 5; waterfowl: χ^2 = 22.3, df = 11, P = 0.02, N = 5; carrion: χ^2 = 55.0, df = 11, P < 0.0001, N = 5). Relative availability of fish was highest in spring during spawning season (Fig. 2.3). Relative availability of waterfowl was highest in summer,

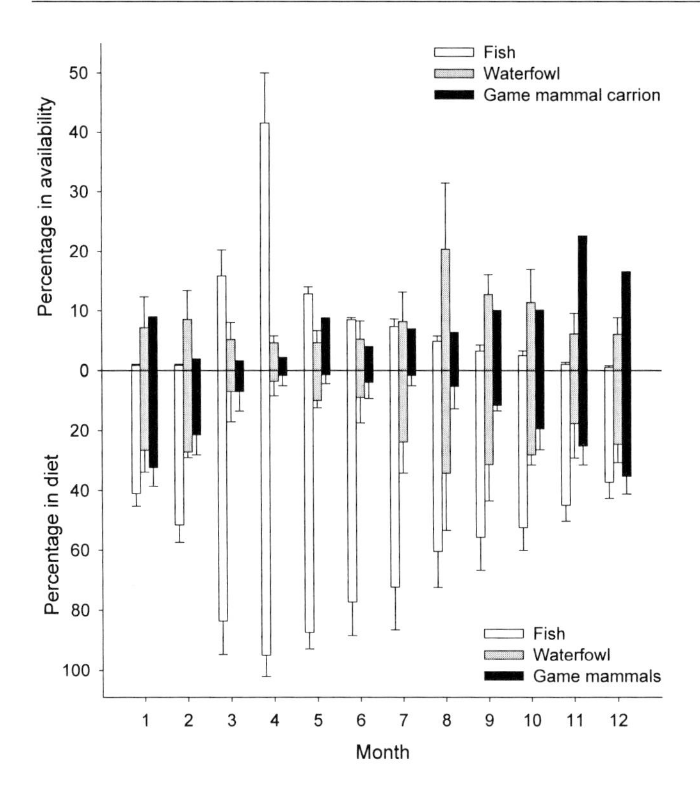

Fig. 2.3. Relationship between seasonal changes in relative food availability and diet of white-tailed eagles. Food availability is expressed as mean monthly percentage of different food types on the total count for each category. The total fish (= 14670 ± 1994) and waterfowl (= 16033 ± 13098) count was sampled within each home range of study pairs SP1 to SP5, the total carrion count (= 2008) within the entire study area. Diet is expressed as mean monthly percentage of different food types in the diet of SP1 to SP5. Error bars denote ± SD.

predominantly consisting of moulting anatids. Relative availability of carcasses of mammals and gut piles was highest in autumn during hunting season. The relative contribution of fish to the diet was best explained by fish availability, of waterfowl to the diet was best explained by fish and waterfowl availability, and of game mammals to the diet was best explained by fish availability (Table 2.3). A monthly increase in the relative contribution of fish to the diet of SP1 to SP5 was significantly positively related to fish availability in their home ranges, but not significantly affected by waterfowl or carrion availability (Table 2.4). The relationship

Table 2.3. *Relative performance of models examining relationship between diet and food availability. For each dependent variable the four candidate models with the lowest Akaike Information Criterion corrected for small sample size (AIC_c) are listed. $\Delta_i \equiv AIC_i - AIC_{min}$; $^{*}P < 0.05$; $^{**}P < 0.01$; $^{***}P < 0.001$. The best-supported models are shown in bold.*

Dependent variable	Candidate models	R^2 (%)	AIC_c	Δ_i
(A) Fish in diet	Fish availability*** + Waterfowl availability + Carrion availability	91.8	-73.9	6.6
	Fish availability*** + Waterfowl availability	92.7	-77.6	2.9
	Fish availability*** + Carrion availability	92.7	-77.6	2.9
	Fish availability***	**93.3**	**-80.5**	**0**
(B) Waterfowl in diet	Fish availability** + Waterfowl availability** + Carrion availability	90.3	-53.8	3.3
	Fish availability*** + Waterfowl availability***	**91.1**	**-57.1**	**0**
	Fish availability* + Carrion availability	57.6	-38.4	18.7
	Fish availability**	61.8	-41.3	15.8
(C) Game mammals in diet	Fish availability** + Waterfowl availability + Carrion availability	75.6	-31.9	4.2
	Fish availability*** + Waterfowl availability	76.7	-33.9	2.2
	Fish availability*** + Carrion availability	76.7	-34.7	1.4
	Fish availability***	**76.3**	**-36.1**	**0**

Table 2.4. *Parameter values for the best-supported models examining the relationship between diet and food availability.*

Dependent variable	F	df	Independent variable	Coefficient ± SE	t	P
(A) Fish in diet	153.1	11	Fish availability	0.969 ± 0.023	12.37	< 0.0001
(B) Waterfowl in diet	57.6	11	Fish availability	-0.616 ± 0.061	-6.44	< 0.0001
			Waterfowl availability	0.558 ± 0.154	5.84	< 0.0001
(C) Game mammals in diet	36.4	11	Fish availability	-0.886 ± 0.148	-6.03	< 0.0001

between their consumption rate of fish and relative fish availability was equivalent to a type II functional response (Fig. 2.4). A monthly increase in the relative contribution of waterfowl to their diet was significantly positively related to waterfowl availability and significantly negatively related to fish availability (Table 2.4). A monthly increase in the relative contribution of game mammals to their diet was significantly negatively related to fish availability. Mean monthly diet breadth of SP1 to SP5 increased as fish availability declined ($\rho = -0.93$, $P < 0.0001$, $N = 12$) and increased with carrion availability ($\rho = 0.63$, $P = 0.028$, $N = 12$).

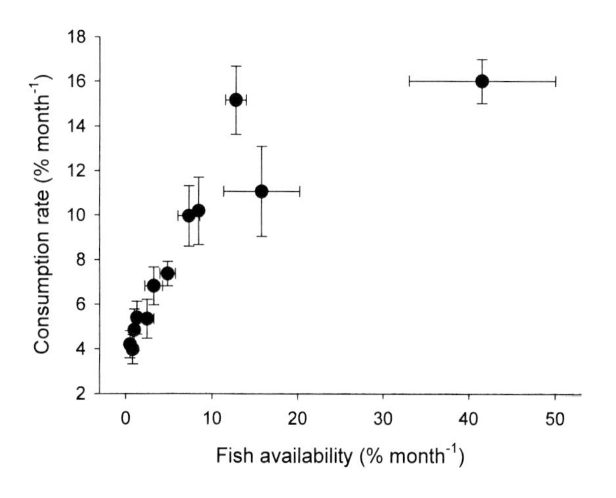

Fig. 2.4. *Relationship between the consumption rate of fish by white-tailed eagles and relative fish availability. The y-axis values represent the mean percentage of fish consumed by the study pairs SP1 to SP5 each month. The x-axis values represent the mean percentage of fish available in the main lakes used for foraging by SP1 to SP5 each month. Error bars denote ± SD.*

Energy intake rate and search effort

The energy intake of white-tailed eagles per unit handling time significantly differed between the food types fish (136.0 ± 25.2 kcal min^{-1}), waterfowl (51.6 ± 13.8 kcal min^{-1}) and carrion (21.2 ± 11.6 kcal min^{-1}; χ^2 = 14.0, df = 2, P = 0.001, N = 7). The net energy intake rate was influenced by the search effort: (1) when the search effort was low, the net energy intake rate was highest for fish and lowest for carrion, (2) when the search effort exceeded a threshold of approximately 15 min, the net energy intake rate was highest for waterfowl and lowest for fish, and (3) when the search effort exceeded a threshold of approximately 30 min, the net energy intake rate was highest for carrion (Fig. 2.5). The search effort of each individual of the study pairs SP1 to SP5 for fish increased significantly in winter half-year (P_{exact} = 0.002, N = 10).

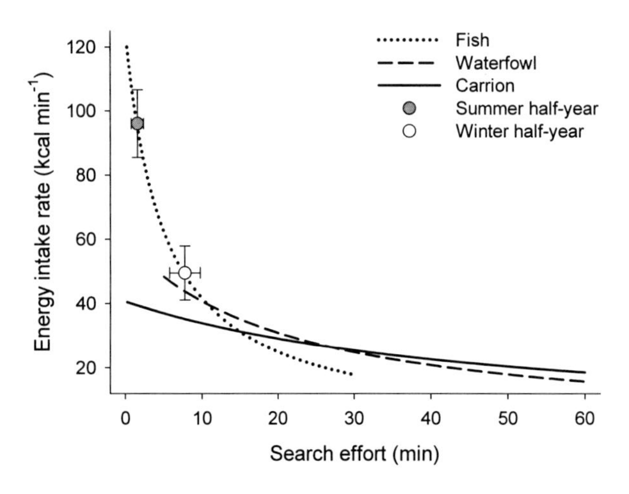

Fig. 2.5. *Energy intake rate (R) of white-tailed eagles as a function of search effort. R = net energy intake / (search effort + handling time). Plot equation based on following data: Mean kilocalories and observed mean handling time and minimum and maximum search effort for fish (N = 100) and waterfowl (N = 25) consumed by the individuals of study pairs SP1 to SP5; mean kilocalories and observed mean handling time (N = 18) and estimated minimum and maximum search effort for carrion consumed by 17 white-tailed eagles in the study area. The grey and white circle represents the mean energy intake rate for fish by the individuals of SP1 to SP5 in the summer half-year and winter half-year, respectively. Error bars denote ± SD.*

Diet and lead poisoning

We compared the monthly consumption of game mammals and waterfowl with the monthly number of eagles (N = 111) found in Germany from 1996 to 2008 and confirmed by necropsy at the IZW to have died from lead poisoning (for further details see Krone *et al.* 2009*a*). The incidence of lead-poisoned eagles varied seasonally and increased with the relative consumption of game mammals, as demonstrated by pellets and food remains from study pairs and stomach contents of dead individuals (study pairs: $\rho = 0.76$, $P = 0.004$, $N = 12$; stomach contents: $\rho = 0.83$, $P = 0.001$, $N = 12$), but was independent of the relative consumption of waterfowl (study pairs: $\rho = 0.13$, $P = 0.685$, $N = 12$; stomach contents: $\rho = 0.51$, $P = 0.089$, $N = 12$).

Discussion

Importance of mammals and carrion in the diet

Our results showed that mammals represent an important food component of white-tailed eagles in autumn and winter. This is in accordance with findings on white-tailed eagles in Lapland (Helander 1983) and on another *Haliaeetus* species, the bald eagle *H. leucocephalus*, in northern Arizona (Grubb & Lopez 2000). Several studies revealed a substantially lower contribution of mammals to the diet of white-tailed eagles. This could be due to (1) focusing exclusively on food items collected during the summer half-year (Wille & Kampp 1983; Sulkava *et al.* 1997), or (2) collecting a considerably smaller number of food remains during the winter half-year by unsystematic sampling (Willgohs 1961; Oehme 1975; Struwe-Juhl 2003). None of these studies had access to GPS-positions of eagles which facilitates the location of carcasses scavenged by eagles. Mammals not only contributed an important part to the diet of all seven study pairs but also to eagles from throughout Germany as documented by the analysis of stomach contents of 126 eagles found dead throughout Germany. For study pair SP7, mammals were the most important diet component. Here, mammals might have been overrepresented, because one GPS-position per day will not allow the detection of every excursion of SP7 to foreign water bodies useful for capturing fish and waterfowl, and the contribution of fish and waterfowl be underrepresented.

Since consumed mammals were predominantly game species with a considerably higher body mass than white-tailed eagles (4.1-6.9 kg; Glutz von Blotzheim *et al.* 1971) and bone remains often exhibited bullet wounds or fractures, they were most likely carrion rather than prey hunted and killed alive (Langgemach & Henne 2001). Since canids such as red foxes and raccoon dogs are smaller but can defend themselves against white-tailed eagles and white-tailed eagles are, unlike golden eagles *Aquila chrysaetos*, lumbering and have poor manoeuvrability (Rudebeck 1950), it is likely that primarily crippled canids were hunted or that they were eaten as carrion. Our findings therefore suggest that most mammals were consumed as carrion, in agreement with other studies documenting regular scavenging on mammal carcasses by white-tailed eagles (Selva *et al.* 2005; M. Nadjafzadeh *et al.* unpublished data).

Functional response to changing food availability

As predicted, white-tailed eagles exhibited a functional response to changes in the availability of fish and waterfowl. Fish seemed to be the primary prey because they were eaten in direct proportion to their availability whereas waterfowl and game mammals were taken in inverse proportion to fish availability, suggesting that they are alternative food components. White-tailed eagles responded to changes in fish availability with a type II functional response (Holling 1959), typical for mammalian and avian predators and their primary prey (Jędrzejewska & Jędrzejewski 1999). Since waterfowl were consumed in proportion to their availability but game mammals were not eaten in relation to carrion availability, waterfowl should be considered preferred food to carrion. Although carrion usually represents the largest food type with highest energy content, followed by waterfowl and then fish (data not shown), such a response appears plausible. Our study pairs spent more time and energy searching for and handling (plucking) waterfowl than for fish. Carrion probably required the greatest search effort because of its unpredictable occurrence in space and time (DeVault *et al.* 2003) and the longest handling time owing to inter- and intraspecific competition (Halley & Gjershaug 1998). Consequently, when fish were abundant and the search effort of the study pairs was low, fish were more profitable than waterfowl or carrion. This trade-off between search effort, handling time and energy intake shifted in autumn and winter in favour of waterfowl and carrion when fish availability sharply decreased but waterfowl and carrion availability increased. Availability of carrion at the end of winter is probably higher than the data from hunting bags indicate because animals that died from natural causes (not included in hunting bags) will become increasingly available, such as ungulates dying from starvation, disease and severe winter conditions (Jędrzejewska & Jędrzejewski 1999).

The functional response is reflected by the result that the diet breadth of all pairs was negatively correlated with fish availability and significantly increased during the winter half-year. It is also consistent with predictions by the alternative prey hypothesis (Angelstam *et al.* 1985) and agrees with other studies where generalist predators responded to a decline in their main food type by switching to alternative food types (Thompson & Colgan 1990; Reif *et al.* 2001).

Dietary shifts in relation to habitat quality

The functional response and the observed diet composition of white-tailed eagles suggest that fish are the most profitable food because they were the main food item and their contribution to the diet increased as diet breadth declined. Birds and mammals/carrion showed a positive correlation with diet breadth and are therefore most likely alternative foods. The wide diet breadth documented by the analysis of stomach contents of eagles from throughout Germany supports the idea that this raptor is a versatile generalist (Sulkava *et al.* 1997). But to what extent and how do white-tailed eagles adjust their foraging to local food supply? Diet varied substantially and significantly between eagles with different causes of death, and diet breadth also varied substantially between study pairs. Study pairs SP4 and SP5 exhibited the narrowest diet breadth. Their home ranges contained one shallow, polytrophic lake each, the main focus of their foraging activities. These lakes were populated by high numbers of *Abramis brama*, a member of the Cyprinidae and preferred prey of white-tailed eagles (M. Nadjafzadeh *et al.* unpublished data). Thus, SP4 and SP5 inhabited home ranges with an ample supply of preferred food, i.e., a high quality habitat. This result is consistent with the prediction by the optimal prey choice model that in a habitat where profitable prey is very common it will not pay the predator to eat prey with low profitability (Stephens & Krebs 1986). The diet breadth of study pairs SP1 to SP3 and SP6 was wider. Their home ranges contained one eutrophic lake each. These lakes were slightly (BS) or considerably (KO) deeper than the water bodies used for foraging by SP4 and SP5. Here, fish were less frequently available close to the water surface than in DS and GS and thus likely to increase foraging effort for eagles, as evidenced by the type II functional response, and stimulating the consumption of alternative food such as preferred, easy to hunt waterfowl such as coots (Struwe-Juhl 2003; M. Nadjafzadeh *et al.* unpublished data). Overall, the home ranges of SP1 to SP3 and SP6 offer an adequate food supply and can therefore be classified as habitats with at least satisfactory quality. In contrast, SP7 exhibited a large diet breadth. Their home range contained only very small, weakly eutrophic lakes unlikely to provide a sufficient food supply of either fish or waterfowl and thus alternative food sources were probably required throughout the year. This would coincide with the prediction that in a poor habitat where profitable prey is scarce and travel time is long, the optimal diet should include more food types than in a good habitat (McArthur & Pianka 1966). Jorde and Lingle (1988) showed that predators may respond to a decline in their main prey species by scavenging and kleptoparasitising other predators. We observed these types of behaviour in SP7 who included a very high proportion of game mammals in their diet and consumed prey species such as oscine birds which they were unlikely to successfully hunt on their own and probably stole from raptors such as peregrine falcons (Oehme 1975). The differing diet breadths of our study pairs indicate that white-tailed eagles pursue individual foraging tactics which take into account local variation in the supply of preferred food items and adjusted diet and diet breadth in a manner consistent with predictions of optimal foraging theory.

Sources of lead fragments

The analyses of stomach contents showed that lead-poisoned eagles fed primarily on ungulate carcasses. These eagles predominantly (at least 75 %) exhibited an acute lead intoxication, i.e., they died within several days of ingesting lead (Pattee *et al.* 1981). One of the clinical signs is gastrointestinal stasis which means that food-pellet formation and egestion no longer take place (Locke & Thomas 1996). Accordingly, the remains found in their stomachs were very likely the source of lead fragments that induced the poisoning. Our result agrees with other studies where deer carcasses and gut piles were considered primary sources of lead for raptors (Hunt *et al.* 2006). It is also consistent with the strong positive correlation between seasonal changes in the consumption of game mammals and the incidence of lead-poisoned eagles. In contrast, waterfowl remains were found only twice in stomachs of lead-poisoned eagles and seasonal changes in the consumption of waterfowl were not correlated with concomitant changes in the incidence of lead-poisoned eagles. This suggests that waterfowl are, if at all, only a minor source of lead fragments and thus of lead poisoning.

Interestingly, train accident victims also predominantly ingested game mammals before they died. These birds fed on mammals killed by trains and lying on the rails and were themselves hit by a train whilst feeding (Krone *et al.* 2003).

Implications for conservation management

We have demonstrated that mammal carcasses and gut piles represent important alternative food sources for white-tailed eagles when more profitable food types such as fish and waterfowl are scarce due to seasonal changes or poor habitat quality. If they are contaminated with lead fragments, they constitute a serious threat for white-tailed eagles and other birds with similar feeding habits. The substantial contribution of large game mammals in terms of carrion to the diet of eagles indicates that a replacement of lead in bullets used in shooting ungulates for hunting is essential to reduce this source of mortality. Conservation management of scavenging birds would therefore be considerably improved if lead did not enter or remain in the carcasses. This suggests that the use of lead-free ammunition would be helpful – provided alternative ammunition to conventional lead ammunition is not toxic in other ways.

This study was initiated and designed by a workshop (Krone & Hofer 2005) to which all relevant stakeholders, including representatives of hunting associations, the ammunition industry, foresters and conservation organisations were invited. One question identified by the stakeholders that needed an answer concerned the main sources of lead poisoning in eagles. We presented progress reports of this study in three annual workshops to the stakeholders as part of the wider research project on causes of and possible solutions to lead poisoning in white-tailed eagles (Krone *et al.* 2009*b*). Currently, the ammunition industry is looking intensively into further developing lead-free alternative bullets, and voluntary use of already available lead-free bullets has started in several federal states in Germany.

Hunters have begun to switch to lead-free ammunition not only in Germany but also in other places such as Hokkaido (Saito 2009) and Arizona as well as California (Sieg *et al.* 2009) where lead ammunition apparently causes a similar problem for the vulnerable Steller's sea eagle (*Haliaeetus pelagicus*) and the critically endangered California condor (*Gymnogyps californianus*). However, since lead poisoning has been documented as a major mortality factor in numerous scavenging birds around the world (Fisher *et al.* 2006), a general changeover from lead-based ammunition to non-toxic alternatives in game mammal hunting need to be implemented for a sustainable improvement of raptor conservation.

Acknowledgements

We are grateful to the administration of the nature park *Nossentiner/Schwinzer Heide* and the *Reepsholt-Stiftung* for their collaboration and logistic support. We thank W. Neubauer for data on waterfowl availability at the *Krakower Obersee*. We are indebted to all cooperating fishermen, forestry districts, hunters and farmers in the study area for collaboration and support, and to F. Scholz, J. Sulawa, A. Trinogga, and N. Kenntner for assistance and support.

This study was funded by the Federal Ministry of Education and Research (BMBF, reference no. 0330720) and the Leibniz Institute for Zoo and Wildlife Research Berlin.

References

Angelstam, P., Lindström, E. & Widén, P. 1985. Synchronous short-term population fluctuations of some birds and mammals in Fennoscandia – occurrence and distribution. *Holarctic Ecology* 8: 285-298.

Bibby, C.J. 1995. *Methoden der Feldornithologie*. Neumann Verlag, Radebeul, Germany.

Burnham, K.P. & Anderson, D.R. 2002. Avoiding pitfalls when using information-theoretic methods. *Journal of Wildlife Management* 66: 912-918.

DeVault, T.L., Rhodes, O.E. & Shivik, J.A. 2003. Scavenging by vertebrates: behavioral, ecological, and evolutionary perspectives on an important energy transfer pathway in terrestrial ecosystems. *Oikos* 102: 224-234.

Fischer, W. 1982. *Die Seeadler*. Ziemsen Verlag, Wittenberg Lutherstadt, Germany.

Fisher, I.J., Pain, D.J. & Thomas, V.G. 2006. A review of lead poisoning from ammunition sources in terrestrial birds. *Biological Conservation* 131: 421-432.

García-Berthou, E. 2001. Size- and depth-dependent variation in habitat and diet of the common carp (*Cyprinus carpio*). *Aquatic Sciences* 63: 466-476.

Glutz von Blotzheim, U.N., Bauer, K.M. & Bezzel, E. 1971. *Handbuch der Vögel Mitteleuropas*. Volume 4: Falconiformes. Akademische Verlagsgesellschaft, Frankfurt am Main, Germany.

Greene, H.W. & Jaksić, F.M. 1983. Food-niche relationships among sympatric predators – effects of level of prey identification. *Oikos* 40: 151-154.

Grubb, T.G. & Lopez, R.G. 2000. Food habits of bald eagles wintering in northern Arizona. *Journal of Raptor Research* 34: 287-292.

Halley, D.J. & Gjershaug, J.O. 1998. Inter- and intra-specific dominance relationships and feeding behaviour of golden eagles *Aquila chrysaetos* and sea eagles *Haliaeetus albicilla* at carcasses. *Ibis* 140: 295-301.

Hanson, K.C., Hasler, C.T., Cooke, S.J., Suski, C.D. & Philipp, D.P. 2008. Intersexual variation in the seasonal behaviour and depth distribution of a freshwater temperate fish, the largemouth bass. *Canadian Journal of Zoology* 86: 801-811.

Hauff, P. 2003. Sea-eagles in Germany and their population growth in the 20th century. In: *Sea eagle 2000*, (eds.) Helander, B., Marquiss, M. & Bowerman, B., pp. 71-77. Swedish Society for Nature Conservation/SNF, Stockholm, Sweden.

Hauff, P. 2008. Seeadler erobert weiteres Terrain. *Nationalatlas aktuell 1 (01/2008)*. Leibniz-Institut für Länderkunde [WWW document], Leipzig, Germany. URL http://nadaktuell.ifl-leipzig.de/Seeadler.1_01-2008.0.html

Helander, B. 1983. Reproduction of the white-tailed sea eagle *Haliaeetus albicilla* (L.) in Sweden, in relation to food and residue levels of organochlorine and mercury compounds in the eggs. *Ph.D. dissertation*, University of Stockholm, Stockholm, Sweden.

Helander, B., Axelsson, J., Borg, H., Holm, K. & Bignert, A. 2009. Ingestion of lead from ammunition and lead concentrations in white-tailed sea eagles (*Haliaeetus albicilla*) in Sweden. *Science of the Total Environment* 407: 5555-5563.

Holling, C.S. 1959. Some characteristics of simple types of predation and parasitism. *Canadian Entomologist* 91: 385-398.

Hunt, W.G., Burnham, W., Parish, C.N., Burnham, B., Mutch, B. & Oaks, J.L. 2006. Bullet fragments in deer remains: implications for lead exposure in scavengers. *Wildlife Society Bulletin* 34: 167-170.

Jędrzejewska, B. & Jędrzejewski, W. 1999. *Predation in vertebrate communities. The Białowieża Primeval Forest as a case study*. Springer Verlag, Berlin, Germany.

Jorde, D.G. & Lingle, G.R. 1988. Kleptoparasitism by bald eagles wintering in south-central Nebraska. *Journal of Field Ornithology* 59: 183-188.

Kenntner, N., Tataruch, F. & Krone, O. 2001. Heavy metals in soft tissue of white-tailed eagles found dead or moribund in Germany and Austria from 1993 to 2000. *Environmental Toxicology and Chemistry* 20: 1831-1837.

Kim, E.Y., Goto, R., Iwata, H., Masuda, Y., Tanabe, S. & Fujita, S. 1999. Preliminary survey of lead poisoning of Steller's sea eagle (*Haliaeetus pelagicus*) and white-tailed sea eagle (*Haliaeetus albicilla*) in Hokkaido, Japan. *Environmental Toxicology and Chemistry* 18: 448-451.

Krebs, C.J. 1999. *Ecological methodology*. 2nd edition. Addison Wesley Longman, New York, USA.

Krone, O., Langgemach, T., Sömmer, P. & Kenntner, N. 2003. Causes of mortality in white-tailed sea eagles from Germany. In: *Sea eagle 2000*, (eds.) Helander, B., Marquiss, M. & Bowerman, B., pp. 211-218. Swedish Society for Nature Conservation/SNF, Stockholm, Sweden.

Krone, O. & Hofer, H. (eds.). 2005. *Bleihaltige Geschosse in der Jagd – Todesursache von Seeadlern*? Leibniz Institute for Zoo and Wildlife Research, Berlin, Germany.

Krone, O., Kenntner, N. & Tataruch, F. 2009*a*. Gefährdungsursachen des Seeadlers (*Haliaeetus albicilla* L. 1758). *Denisia* 27: 139-146.

Krone, O., Kenntner, N., Trinogga, A., Nadjafzadeh, M., Scholz, F., Sulawa, J., Totschek, K., Schuck-Wersig, P. & Zieschank, R. 2009*b*. Lead poisoning in white-tailed sea eagles: causes and approaches to solutions in Germany. In: *Ingestion of lead from spent ammunition: implications for wildlife and humans*, (eds.) Watson, R.T., Fuller, M., Pokras, M. & Hunt, G., pp. 289-301. The Peregrine Fund, Idaho, USA.

Langgemach, T. & Henne, E. 2001. Storks *Ciconia nigra*, *C. ciconia* and common cranes *Grus grus* as prey of the white-tailed eagle *Haliaeetus albicilla*. *Vogelwelt* 122: 81-87.

Locke, L.N. & Thomas, N.J. 1996. Lead poisoning of waterfowl and raptors. In: *Noninfectious diseases of wildlife*, (eds.) Fairbrother, A., Locke, L.N. & Hoff, G.L., pp. 108-117. Iowa State University Press, Iowa, USA.

Martin, P. & Bateson, P. 1993. *Measuring behaviour*. 2nd edition. Cambridge University Press, Cambridge, UK.

Martina, P.A., Campbell, D., Hughes, K. & McDaniel, T. 2008. Lead in the tissues of terrestrial raptors in southern Ontario, Canada, 1995-2001. *Science of the Total Environment* 391: 96-103.

McArthur, R.H. & Pianka, E. 1966. On optimal use of a patchy environment. *American Naturalist* 100: 603-609.

Meretsky, V.J., Snyder, N.F.R., Beissinger, S.R., Clendenen, D.A. & Wiley, J.W. 2000. Demography of the California condor: implications for reestablishment. *Conservation Biology* 14: 957-967.

Mersmann, T.J., Buehler, D.A., Fraser, J.D. & Seegar, J.K.D. 1992. Assessing bias in studies of bald eagle food habits. *Journal of Wildlife Management* 56: 73-78.

Miller, M.J.R., Wayland, M.E. & Bortolotti, G.R. 2002. Lead exposure and poisoning in diurnal raptors: a global perspective. In: *Raptors in the New Millenium*, (eds.) Yosef, R., Miller, M.L. & Pepler, D., pp. 224-245. International Birding and Research Center, Eilat, Israel.

Oehme, G. 1975. Ernährungsökologie des Seeadlers, *Haliaeetus albicilla* (L.), unter besonderer Berücksichtigung der Population in den drei Nordbezirken der Deutschen Demokratischen Republik. *Doctoral dissertation*, Universität Greifswald, Greifswald, Germany.

Pain, D.J. & Amiard-Triquet, C. 1993. Lead poisoning of raptors in France and elsewhere. *Ecotoxicology and Environmental Safety* 25: 183-192.

Pattee, O.H., Wiemeyer, S.N., Mulhern, B.M., Sileo, L. & Carpenter, J.W. 1981. Experimental lead-shot poisoning in bald eagles. *Journal of Wildlife Management* 45: 806-810.

Pyke, G.H., Pulliam, H.R. & Charnov, E.L. 1977. Optimal foraging: a selective review of theory and tests. *The Quarterly Review of Biology* 52: 137-154.

Reif, V., Tornberg, R., Jungell, S. & Korpimäki, E. 2001. Diet variation of common buzzards in Finland supports the alternative prey hypothesis. *Ecography* 24: 267-274.

Rudebeck, G. 1950. The choice of prey and modes of hunting of predatory birds with special reference to their selective effect. *Oikos* 2: 65-88.

Saito, K. 2009. Lead poisoning of Steller's sea eagle (*Haliaeetus pelagicus*) and white-tailed eagle (*Haliaeetus albicilla*) caused by the ingestion of lead bullets and slugs, in Hokkaido, Japan. In: *Ingestion of lead from spent ammunition: implications for wildlife and humans*, (eds.) Watson, R.T., Fuller, M., Pokras, M. & Hunt, G., pp. 302-309. The Peregrine Fund, Idaho, USA.

Scheuhammer, A.M. & Templeton, D.M. 1998. Use of stable isotope ratios to distinguish sources of lead exposure in wild birds. *Ecotoxicology* 7: 37-42.

Schoener, T.W. 1971. Theory of feeding strategies. *Annual Review of Ecology and Systematics* 2: 369-404.

Scholz, F. 2010. Spatial use and habitat selection of white-tailed eagles (*Haliaeetus albicilla*) in Germany. *Doctoral dissertation*, Freie Universität Berlin, Berlin, Germany.

Selva, N., Jędrzejewska, B. & Jędrzejewski, W. 2005. Factors affecting carcass use by a guild of scavengers in European temperate woodland. *Canadian Journal of Zoology* 83: 1590-1601.

Sieg, R., Sullivan, K.A. & Parish, C.N. 2009. Voluntary lead reduction efforts within the northern Arizona range of the California condor. In: *Ingestion of lead from spent ammunition: implications for wildlife and humans*, (eds.) Watson, R.T., Fuller, M., Pokras, M. & Hunt, G., pp. 341-349. The Peregrine Fund, Idaho, USA.

Sinclair, A.R.E., Freyxell, J.M. & Caughley, G. 2006. Wildlife ecology, conservation, and management. 2nd edition. Blackwell Science, Oxford, UK.

Sonerud, G.A. 1992. Functional response of birds of prey – biases due to the load-size effect in central place foragers. *Oikos* 63: 223-232.

Stephens, D.W. & Krebs, J.R. 1986. *Foraging theory*. Princeton University Press, Princeton, USA.

Struwe-Juhl, B. 2003. Why do white-tailed eagles prefer coots? In: *Sea eagle 2000*, (eds.) Helander, B., Marquiss, M. & Bowerman, B., pp. 317-326. Swedish Society for Nature Conservation/SNF, Stockholm, Sweden.

Sulawa, J., Robert, A., Köppen, U., Hauff, P. & Krone, O. 2010. Recovery dynamics and viability of the white-tailed eagle (*Haliaeetus albicilla*) in Germany. *Biodiversity and Conservation* 19: 97-112.

Sulkava, S., Tornberg, R. & Koivusaari, J. 1997. Diet of the white-tailed eagle *Haliaeetus albicilla* in Finland. *Ornis Fennica* 74: 65-78.

Teerink, B.J. 1991. *Hair of west European mammals*. Cambridge University Press, Cambridge, UK.

Thomas, V.G. & Owen, M. 1996. Preventing lead toxicosis of European waterfowl by regulatory and nonregulatory means. *Environmental Conservation* 23: 358-364.

Thomson, I.D. & Colgan, P.W. 1990. Prey choice by marten during a decline in prey abundance. *Oecologia* 83: 443-451.

Wille, F. & Kampp, K. 1983. Food of the white-tailed eagle *Haliaeetus albicilla* in Greenland. *Holarctic Ecology* 6: 81-88.

Willgohs, J.F. 1961. The white-tailed eagle *Haliaeetus albicilla albicilla* (Linne) in Norway. *Ph.D. dissertation*, University of Bergen, Bergen, Norway.

CHAPTER 3

Sit-and-wait for large prey: foraging strategy and prey choice of white-tailed eagles in northeastern Germany

Abstract

Little is known about foraging strategy and prey choice in large raptor species such as the white-tailed eagle *Haliaeetus albicilla* and how they might change with age and season. Here, we present results about time allocation, foraging pattern and diet selection of adult territorial white-tailed eagles from northeastern Germany. To assess age-related differences, we also observed foraging behaviour in roaming juveniles. White-tailed eagles allocated most of their diurnal time to perching, which implies an energy maximising rather than time minimising foraging strategy. Since perch-hunting was more efficient than flight-hunting, "sit-and-wait" for prey seems to be a low-cost, highly profitable foraging mode in eagles. A linear mixed model revealed that season significantly affected eagle foraging patterns. Success in prey capture decreased and duration of foraging flights increased considerably in winter. Eagle strike success significantly varied between different territories and increased with increasing habitat quality in terms of profitable hunting grounds. Adult eagles foraged more efficiently than juveniles, presumably because of their superior spatial knowledge and hunting skills. A use-availability design for prey selectivity indices judged by log-likelihood chi-square statistics indicated that eagles make choices, both within their primary prey fish and their alternative prey waterfowl, consistent with predictions of optimal diet models. Provided that prey was abundant, eagles preferred large over small fish and slow over agile waterfowl species. Eagles foraged opportunistically for game carrion probably because of its ephemeral nature and patchy distribution. Thus, prey choice by white-tailed eagles reflected a complex function of absolute availability, size and anti-predator behaviour of their prey. Our study demonstrates that white-tailed eagles are generally energy maximisers and pursue a "sit-and-wait" hunting mode to capture profitable prey, and can modify their foraging strategy to cope with variations in weather conditions and food availability.

Keywords: dietary preferences, energy maximisers, food availability, foraging behaviour, *Haliaeetus albicilla*, prey size selection, time allocation

Introduction

A central issue in behavioural ecology is the allocation of time and energy to different activities (Masman *et al.* 1988). Foraging differs from other activities by its exclusive quality of providing the individual with a net energy profit (Lundberg 1985). Therefore, much attention has been paid to the time expended on foraging in the total time budget (Pyke *et al.* 1977; Lemon 1991; Shepard *et al.* 2009). In terms of optimality models relating foraging time to potential fitness, animals are thought of as either "time minimisers", who maximise fitness by minimising time spent foraging to gather a given amount of energy, or "energy maximisers", who maximise fitness by maximising net energy intake for a given time spent foraging (Schoener 1971). In reality, these foraging strategies describe the extremes of a continuum of options which may be influenced by factors such as environmental conditions (Edwards 1997), predation risk (Oksanen & Lundberg 1995) and individual variation (Mikheev & Wanzenböck 1999). Depending on the adopted foraging strategy, predators use different foraging modes such as "actively foraging" which would be expected by time minimisers and "sit-and-wait" which would be applied by energy maximisers (Perry 1999; Jackson *et al.* 2004).

The predictions of the theory of optimal foraging have been commonly tested with small to medium-sized birds (e.g., Hixon & Carpenter 1988; Jenkins 2000; van Gils *et al.* 2005), but rarely with large raptor species (but see Bakaloudis 2010). The white-tailed eagle *Haliaeetus albicilla* is one of the largest European raptors. Empirical data on its foraging strategy and hunting mode are currently lacking. Since wing loading (body mass divided by wing area) is negatively related to body size and foraging flight activity (Jaksić & Carothers 1985; Gamauf 1998), we would expect white-tailed eagles to be active searchers. Contrary to this, the similar-sized bald eagle (*Haliaeetus leucephalus*) has been observed to allocate most of its time to perching (Watson *et al.* 1991), indicating a passive search mode. Accordingly, both active foraging and sit-and-wait are conceivable foraging strategies of white-tailed eagles.

In most species, availability of food resources has been identified as the key factor that shapes foraging behaviour and dietary decisions (Stephens & Krebs 1986; Naef-Daenzer *et al.* 2000). It is supposed that a species will select resources that are most suitable for its life requirements, and that high quality resources are selected over low quality ones (Manly *et al.* 2002). The main food resources of white-tailed eagles are fish, birds, and carrion (Oehme 1975; Helander 1983; Sulkava *et al.* 1997). We have already demonstrated that the species prefers fish and uses waterfowl and carcasses of game mammals as alternative food types (M.

Nadjafzadeh *et al.* unpublished data). However, white-tailed eagles are known to consume a remarkably wide variety of prey species and information about prey choice beyond prey class level is scarce (but see Struwe-Juhl 2003). We hypothesise that despite of their wide prey spectrum white-tailed eagles distinguish between available species of differing profitability and select the more profitable ones, as predicted by the optimal prey choice model (Krebs 1978). Furthermore, since it is generally assumed that large predators select large prey over small food items (Goss-Custard 1977; Karanth & Sunquist 1995; Gil & Pleguezuelos 2001), we expect diet selection in the white-tailed eagle to be size-dependent.

An individual's foraging behaviour and prey choice has implications for its reproductive output, survival, expansion, and, ultimately, the growth rate of the population (Sinclair & Krebs 2002). Hence, understanding dietary demands of white-tailed eagles is not only a step towards understanding the interaction between a large raptor and its environment, but also a step towards its conservation. This species was almost eradicated from many European countries owing to human persecution (Fischer 1982; Thiollay 1994) and, despite of the recovery of white-tailed eagle populations in northern and central Europe during the last two decades (Helander & Sternberg 2002; Hauff 2003), it still suffers from anthropogenic threats such as lead poisoning, collisions with trains and wires, and electrocution (Krone *et al.* 2009*a*).

In this paper, we investigate the foraging pattern and diet selection of white-tailed eagles. Our specific objectives were to (1) examine the allocation of time to various daily activities to determine foraging strategy and hunting mode, (2) identify aspects of the environment and life history of eagles that affect their foraging behaviour, (3) test whether eagles display dietary preferences for particular species or prey sizes, and (4) assess the implications of our data for prospective conservation management of white-tailed eagles.

Material and methods

Study site and study animals

The study was conducted in the nature park *Nossentiner/Schwinzer Heide* in northeastern Germany (53°30'-53°40'N, 11°59'-12°35'E), between January 2007 and December 2008. This bird sanctuary is part of the Mecklenburg Lake District which hosts the core of the German white-tailed eagle population (Hauff 2003). The area is characterised by numerous freshwater lakes, large forests (dominated by *Pinus sylvestris*), extensive pastures and low

Table 3.1. *White-tailed eagle study pairs (SP) and lakes most frequently used for foraging activities. SP1, SP2 and SP3 inhabited the eastern, western and northern part of the Krakower Obersee, respectively.*

Study pairs (SP)	Main water body for eagle foraging	Area (ha)	Maximum depth (m)	Mean depth (m)	Trophic state
SP1 SP2 SP3	*Krakower Obersee* (KO)	799	28.3	7.5	eutrophic
SP4	*Damerower See* (DS)	285	7.0	2.0	polytrophic
SP5	*Goldberger See* (GS)	770	4.1	2.1	polytrophic

human population density. Study animals were five territorial white-tailed eagle pairs (SP1 to SP5, Table 3.1). The females of pairs SP1 and SP3 and the male of SP4 were fitted with a GPS-transmitter to examine space and habitat use (Scholz 2010). Each study pair's home range contained a larger water body (≥ 285 ha) on which the eagles focused their foraging. In addition, we observed juvenile eagles that inhabited temporarily the territories of the study pairs.

Data collection

Time allocation and foraging behaviour

To investigate diurnal and seasonal patterns of foraging activity and prey captures, we observed each study pair in its foraging area on two days per month. For observations we chose in each home range one vantage point which offered the best view (at least 2 km) over the main water bodies used by the study pairs during foraging. Our observations were aided by 10 × 42 binoculars with integrated laser rangefinder (Geovid HD 42, Leica Camera AG, Germany) and a 20-60 × spotting scope (Diascope 85T*FL, Carl Zeiss AG, Germany). We used focal animal sampling combined with continuous recording (Martin & Bateson 1993) to record foraging activities of each individual of the study pairs from dawn to dusk. We distinguished individual focal eagles by body size and plumage characteristics. The following behavioural categories were recorded: perching, sitting, directional locomotion, soaring flight, foraging flight, feeding, and other (see Table 3.2 for detailed definitions of each category). It was not possible to discriminate unambiguously among the presence or absence of foraging "intentions" of a perching bird. We defined a flight as 'foraging flight' when the focal animal

Table 3.2. *Description of behavioural categories used in focal animal sampling protocols to record time budgets of ten territorial white-tailed eagles.*

Behavioural category	Description
Perching (P)	sitting with view onto the lake, grassland etc.
Sitting (S)	sitting without view onto the lake, grassland etc., including sitting on nest
Directional locomotion (DL)	walking, running, flying with substantial change of location a necessary consequence
Soaring flight (SF)	gliding, soaring, thermal circling; substantial change of location not a necessary consequence
Foraging flight (FF)	flying close to water surface with view onto the lake, circling or hovering over potential prey; substantial change of location not a necessary consequence
Feeding (F)	eating and manipulating food
Other (O)	bathing, drinking, nest-building, social interactions, aggressive interactions

showed a clearly visible foraging behaviour such as flying close to the water surface or circling/hovering over potential prey. Soaring flights do not come into this category but can also be used by eagles to search for prey. To assess the importance of perching and soaring for eagle foraging behaviour, we recorded for every strike whether it was launched directly from the perch, occurred after soaring or ended foraging flights.

To detect age-related differences in foraging behaviour, we also observed, at the observation days in each study pair home range, juvenile eagles. Owing to dispersal and nomadic behaviour, the number of juveniles in each study pair home range fluctuated highly and their length of stay was very short. Thus, long-term observations on individual juveniles, as in case of the territorial adult focal eagles, were not possible. We used therefore behaviour sampling combined with continuous recording (Martin & Bateson 1993) to note opportunistically each visible foraging flight and prey capture attempt of juvenile eagles.

For each observation of foraging eagles, we recorded the date, lake, and time of day. We measured ambient temperature (in °C) and wind speed (in km^{-1}) with a hand-held thermometer-anemometer, SM-18 Skymate (Weathertec GmbH, Burgdorf, Germany). When a prey strike occurred, we assessed the taxonomic classification of the prey individual to at least family level. When the strike occurred on the lake, we also recorded whether it was within the littoral zone (i.e. < 100 m to shore) or open water (i.e. ≥ 100 m to shore). Finally, if the strike resulted from perch-hunting, we documented which kind of perch was used (tree, shrub, etc.) and its distance to the shore.

Diet selection

In studies of resource selection, the first step is to determine the scale of selection to focus on (Manly *et al.* 2002). Since we wished to detect diet preferences in territorial animals, we measured food resource use for each study pair and food availability in each home range, corresponding to study design III defined by Thomas and Taylor (1990) which accounts for a territory-specific food base. To obtain representative results for diet selection throughout the year, we collected data monthly over ten days per month.

To examine food resource use, we systematically searched the home ranges of each study pair, aided by GPS-positions and feeding sites and roosts known from direct observations, for food remains and pellets. Except for breeding season (February-May), we also explored the vicinity of eyries of all study pairs. We ringed nestlings of study pairs and included fresh food items or pellets from eyries in our sample.

To measure food availability, we used random sampling. We defined relative food availability as the abundance of the main food types in areas accessible to eagles. The main food types of our study pairs were fish, waterfowl, and game mammal carcasses (M. Nadjafzadeh *et al.* unpublished data). We assessed the relative abundance of fish in water depths accessible to eagles (approximately 0-1 m; Fischer 1982) in the main water bodies where we observed eagle foraging, the *Krakower Obersee* (KO), *Damerower See* (DS) and *Goldberger See* (GS), in collaboration with local fishermen. We counted the contents of their fish traps (12 to 15 in each lake, distributed around the lake in proximity to the shore) and classified their catch qualitatively by species and quantitatively in terms of number and size. We recorded the relative abundance of waterfowl on the part of the lake surfaces accessible to eagles (Oehme 1975) and used by eagles for foraging by direct observations (Bibby 1995): we counted all waterfowl visible on KO, DS, and GS by using two to six locations as simultaneous recording stations, respectively. We estimated the relative abundance of mammal carcasses and gut piles left by hunters in the field and accessible to eagles (Hunt *et al.* 2006) from hunting bags of the home ranges of all study pairs. These data were provided by the resident forestry districts and the hunters in charge of local hunting blocks. It was not possible to obtain separate hunting bags for each home range. Nevertheless, since the home range of each study pair contained a considerable segment of forest (Scholz 2010) used by hunters, it is most likely that a similar proportion of carcasses and gut piles was available to eagles in each home range.

Data analysis

Time allocation and foraging behaviour

Time allocation and foraging behaviour were analysed on a yearly basis and separately for each month or season. Seasons were defined both as winter (December-February), spring (March-May), summer (June-August) and autumn (September-November), and as winter half-year (September-February) and summer half-year (March-August). Daily variation in foraging flights was divided into three periods: morning (dawn to 11:00$_{AM}$), midday (11:00$_{AM}$ to 3:00$_{PM}$) and afternoon (3:00$_{PM}$ to dusk).

The relative allocation of time of each individual of the study pairs SP1 to SP5 to different activities was calculated by dividing the duration of activity *i* of eagle *j* by the total duration of activities of eagle *j*. The relative strike success of each focal eagle was measured by dividing the number of successful strikes of eagle *j* by the total number of strikes of eagle *j*. We also calculated the strike success of each focal eagle for different prey classes, hunting modes, and seasons separately. We estimated the relative strike success for juvenile eagles at each lake by dividing the number of successful strikes of juvenile eagles at lake *l* by the total number of strikes of juvenile eagles at lake *l*.

Diet selection

For taxonomic identification of collected food items and pellet contents, we prepared a reference collection of bones, scales, feathers and hairs. We also used reference material of the Berlin Museum of Natural History and published identification keys (Teerink 1991). The size of consumed fish species was back-calculated on the basis of recorded scale annuli and radii and the measure of head skeleton bones (Fahmy 1982; Tarkan *et al.* 2007; H. Winkler unpublished data). We computed food resource use and food availability within main food types (i.e. fish, waterfowl, game mammals) as follows: The relative contribution of food species to the diet of each study pair was calculated by dividing the number of species *i* of food type *p* by the total number of collected food items of food type *p*. The relative availability of food species was estimated for the home range of each study pair by dividing the sample count of species *i* of food type *p* by the total sample count of food type *p*.

To test the hypothesis that white-tailed eagles select between prey species, we compared the proportion of different prey species in the diet of eagle pairs with the proportion of different prey species available in their home ranges according to the use-availability approach suggested by Many *et al.* (2002). We calculated the selection ratio, separately for each food type, of each study pair for the *i*th food species as

$$\hat{w}_{ij} = (u_{ij} / u_{i+}) / \pi_{ij}$$

where u_{ij} is the number of food items of species i consumed by study pair j, u_{i+} is the number of food items of species i consumed by all study pairs, and π_{ij} is the proportion of food items of species i available to study pair j. The selection ratio $\hat{w}_{ij}$ can range from 0 to ∞, with values < 1 indicating consumption less frequently than expected (avoidance) and values > 1 indicating consumption more frequently than expected (preference) from the species' respective availabilities.

We tested for prey size selection in white-tailed eagles on the basis of differing size classes of their primary prey – fish (M. Nadjafzadeh *et al.* unpublished data). We compared the proportion of different size classes of fish species in the diet of each eagle pair with the proportion of different size classes available in the water bodies of each home range by calculating the selection ratio $\hat{w}_{ij}$. When the food species and/or fish size selection did not significantly differ between the study pairs SP1, SP2 and SP3, which focused their foraging on the same water body (KO), we combined data from these eagle pairs for further comparison with SP4 and SP5.

Statistical analysis

All statistical tests were two-tailed and carried out with SPSS 16.0 (SPSS Inc., Chicago, IL, USA). The significance level α was set at $P < 0.05$. Unless otherwise indicated, data are presented as mean ± standard deviation (SD). In order to assess differences within time allocation, hunting mode and prey capture behaviour of the focal eagles, we employed Wilcoxon signed-ranks and Friedman tests. We used log-likelihood chi-square tests to compare variations within relative frequency of strike success between lakes, sexes, age classes (adult, juvenile), and study pairs. We examined whether the foraging flight duration (FFD) was affected by environmental factors such as wind speed, temperature, period of day and season, and individual factors such as age and sex, by fitting a linear mixed model using restricted maximum-likelihood analysis (REML). We included the individuals nested within lakes as random factors in the REML to control for the dependence of observations by using the Wald statistic. We log-transformed the values of the dependent variable to ensure that there were no significant deviations of the residuals from normality as judged by the Lilliefors version of the Kolmogorov-Smirnov test. Following Manly *et al.* (2002), we adopted log-likelihood chi-square statistics to measure (1) the extent to which the population of eagles and/or individual eagle pairs were on average using respective food resources in proportion to availability and (2) whether the study pairs of KO were using resources in a similar way. Furthermore, the selection ratios $\hat{w}_{ij}$ of every species and prey size category i were tested for significant deviation from 1 by comparing

$$\{(\hat{w}_{ij} - 1) / \mathrm{SE}(\hat{w}_{ij})\}^2$$

with critical values for the χ^2 distribution with one degree of freedom. We used Bonferroni-adjusted significance levels to account for an increased risk of type I error during multiple comparisons (Manly *et al.* 2002).

Results

Time budget and foraging duration

A total of 993 h of observation revealed the following diurnal time-activity budget for ten territorial white-tailed eagles: 81.5 ± 3.3 % perching, 10.7 ± 2.3 % sitting, 6.8 ± 3.0 % locomotor activity, 0.9 ± 0.2 % feeding, and 0.1 ± 0.1 % other. The proportion of each activity category did not vary significantly among seasons except locomotor activity, which increased in winter half-year (Wilcoxon signed-ranks test, $P_{exact} = 0.014$, $N = 10$). Within locomotor activity, eagles allocated significantly more time to soaring (45.8 ± 10.5 %) and foraging flights (45.1 ± 11.0 %) than directional locomotion (9.1 ± 1.9 %; Friedman test, $\chi^2 = 15.8$, $df = 2$, $P < 0.0001$, $N = 10$).

Overall, we observed 1,395 foraging flights. The foraging flight duration (FFD) ranged between 0.2 and 30 min, with a mean duration of 2.2 ± 2.4 min. Variance in FFD from observations on different individuals nested within lakes was not significant (Wald statistic, $Z = 0.75$, $df = 13$, $P = 0.46$). The FFD decreased as temperature increased (REML, $F_{1,1376.0} = 5.49$, $P = 0.019$; Fig. 3.1A), but was independent from wind speed ($F_{1,1360.2} = 3.76$, $P = 0.053$; Fig. 3.1B). FFD significantly increased from spring and summer to autumn and winter months ($F_{11,1364.8} = 26.7$, $P < 0.0001$; Fig. 3.1C) and was longest at midday compared to morning or afternoon ($F_{2,1374.2} = 4.21$, $P = 0.015$; Fig. 3.1D). FFD significantly decreased from juveniles to adults ($F_{1,978.1} = 23.1$, $P < 0.0001$; Fig. 3.1E), but was independent from sex ($F_{2,20.1} = 1.16$, $P = 0.33$; Fig. 3.1F).

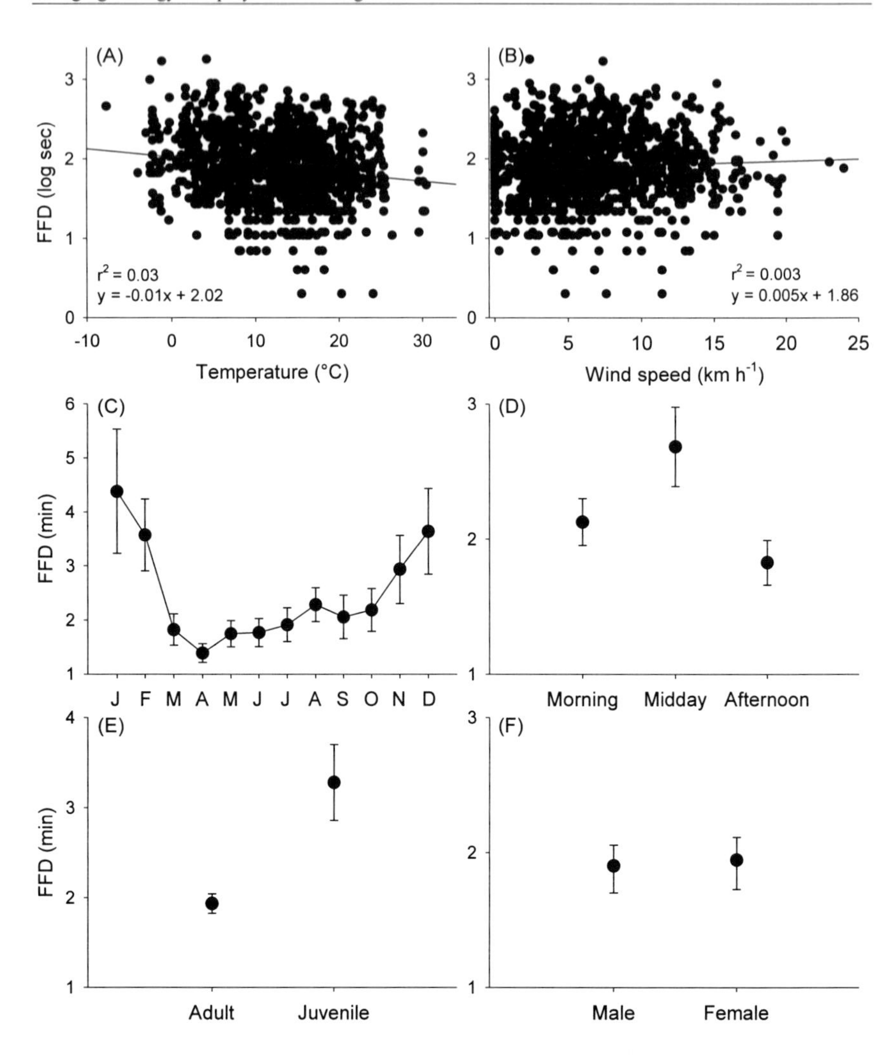

Fig. 3.1. *Variation in foraging flight duration (FFD) of white-tailed eagles in northeastern Germany. Relationship between FFD and (A) temperature and (B) wind speed, with fitted regression lines. (C) Monthly, (D) daily, (E) age-related, and (F) sex-related variation in FFD. Data are expressed as single values of the duration of observed foraging flights (N = 1395; A, B) and as mean duration of foraging flights per respective categories (C-F). The influences of wind speed and sex on FFD were not significant. Error bars represent 95 % confidence intervals.*

Prey capture behaviour and success

We observed 327 prey capture attempts and 165 successful prey captures (overall relative strike success = 50.5 %). Strike success did not differ significantly among sexes (χ^2 log-likelihood test, $\chi^2 = 0.001$, df = 1, P = 0.98, $N_{male\ strikes} = 125$, $N_{female\ strikes} = 132$; Fig. 3.2A), but was higher in adults than juveniles ($\chi^2 = 5.07$, df = 1, P = 0.02, $N_{adult\ strikes} = 257$, $N_{juvenile\ strikes} = 70$; Fig. 3.2B). We also found significant differences in relative strike success between lakes ($\chi^2 = 7.24$, df = 2, P = 0.03, $N_{strikes} = 327$; Fig. 3.2C) and study pairs ($\chi^2 = 10.2$, df = 4, P = 0.04, $N_{strikes} = 257$). Strike success of study pairs SP1 to SP3 (KO) was with 41.4 ± 3 % very similar, and distinctively lower than the relative strike success of SP4 (64.9 %; DS) and SP5 (57 %; GS). Prey targeted by the study eagles consisted of 77.0 ± 16.0 % fish, 21.1 ± 14.5 % waterfowl, and 1.9 ± 4.9 % mammals (hares). Strike success differed significantly among target prey (Friedman test, $\chi^2 = 18.6$, df = 2, P < 0.0001, N = 10) and was highest for fish and lowest for mammals (Fig. 3.2E). Season significantly affected strike success of each study eagle ($\chi^2 = 25.5$, df = 3, P < 0.0001, N = 10) which was highest in spring (78.1 ± 20 %) and lowest in winter (15.1 ± 13.1 %; Fig. 3.2F). Although absolute number of strikes of territorial individuals were higher for flight-hunting (62.1 ± 10.1 %) than perch-hunting (37.9 ± 10.1 %; Wilcoxon signed-ranks test, $P_{exact} = 0.004$, N = 10), relative strike success was significantly higher in perch-hunting than flight-hunting ($P_{exact} = 0.049$, N = 10; Fig. 3.2D). Within flight-hunting, strikes occurred predominantly after foraging flights (92.0 ± 7.2 %) and rarely after soaring flights (8.0 ± 7.2 %; $P_{exact} = 0.002$, N = 10). Perches from which strikes were conducted were predominantly trees (88.8 ± 17.7 %) and occasionally shrubs (6.1 ± 9.6 %) and grass mounds (5.1 ± 10.6 %). The shore distance of perches used by the study eagles ranged from 0 to 100 m, with a mean of 15.6 ± 12.7 m. Strikes did not occur at random lake locations but with 80.3 ± 11.3 % predominantly within the littoral zone ($P_{exact} = 0.002$, N = 10).

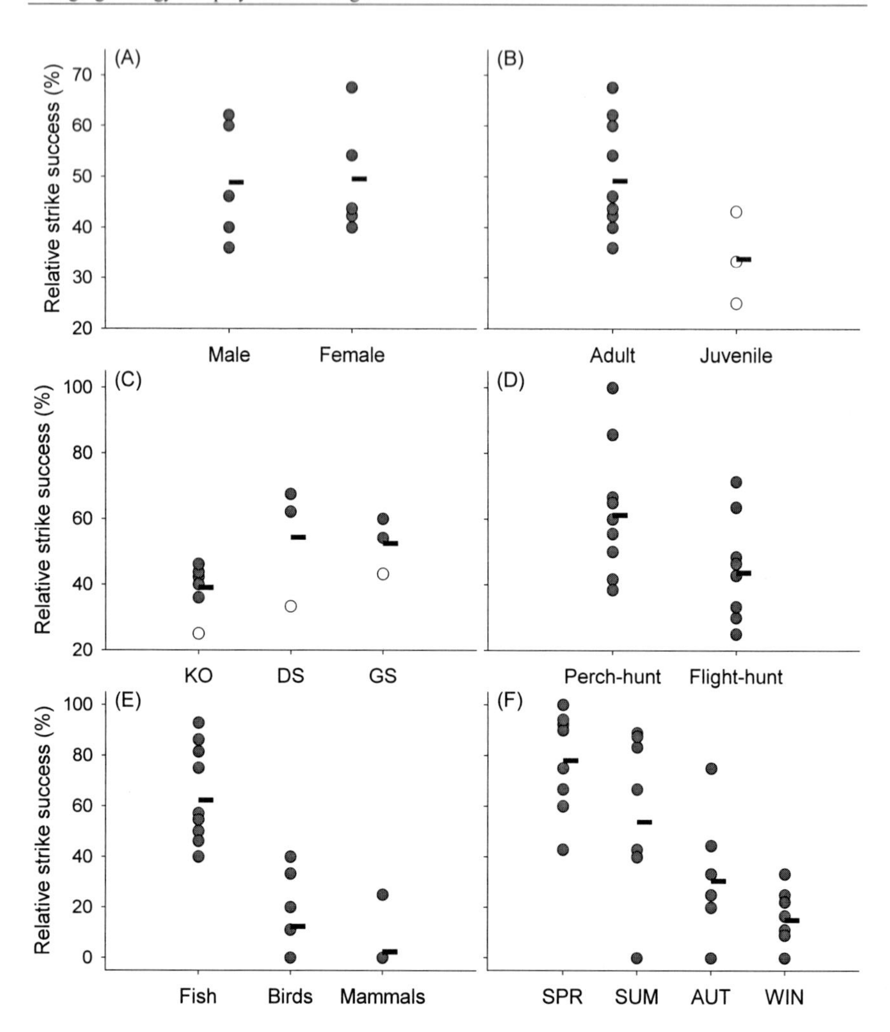

Fig. 3.2. *Variation in relative strike success of white-tailed eagles in northeastern Germany. (A) Sex-related, (B) age-related, (C) lake-related, (D) hunting mode-related, (E) prey-related, and (F) season-related variation in relative strike success. Data are expressed as mean strike success of individual adults (grey dots), juveniles of each lake (white dots), and overall means (black stripes). Only differences in strike success among sexes were not significant. Error bars represent ± 1 SD.*

Diet selection by species

Results of food species resource use and availability are summarised in Table 3.3. The proportion of different fish species in the diet of eagles did not reflect their respective availabilities, both on a population level (χ^2 log-likelihood test, $\chi^2 = 443.7$, df = 6, P < 0.0001) and in each study pair ($\chi^2 \geq 35.8$, df = 6, P < 0.0001). Since the study pairs SP1 to SP3 at the *Krakower Obersee* (KO) selected food species in a similar way ($\chi^2 \leq 12.1$, d.f. $\geq$ 8, P $\geq$ 0.60), we combined their data in the species selection analysis. European eel (*Anguilla anguilla*) was apparently not consumed by eagles although available in low densities. SP1 to SP3 consumed significantly higher proportions of northern pike (*Esox lucius*) than expected from its availability. All study pairs highly preferred carp bream (*Abramis brama*) and avoided common roach (*Rutilus rutilus*). European perch (*Perca fluviatilis*) was avoided by four study pairs and never consumed by SP4. Common rudd (*Scardinius erythrophthalmus*) and white bream (*Blicca bjoerkna*) were consumed according to availability.

Within waterfowl species, there was a selection on the population level ($\chi^2 = 19.7$, df = 7, P = 0.006) and in pairs SP1 to SP3 ($\chi^2 = 75.3$, df = 7, P < 0.0001), but not in SP4 and SP5 ($\chi^2 \leq 7.80$, df = 7, P $\geq$ 0.35), which inhabited home ranges of comparatively low waterfowl availability. The study pairs of the KO showed a strong preference for Eurasian coot (*Fulica atra*), a species that did not occur at the DS and GS, and avoidance of diving duck (*Aythya* spec.). The selection ratios for the remaining waterfowl species great crested grebe (*Podiceps cristatus*), great cormorant (*Phalacocorax carbo*), mute swan (*Cygnus olor*), greylag goose (*Anser anser*), dabbling duck (*Anas* spec.), and goosander (*Mergus merganser*) showed no significant deviation from 1.

Carcasses of the game mammal species fallow deer (*Cervus dama*), roe deer (*Capreolus capreolus*), wild boar (*Sus scrofa*), red fox (*Vulpes vulpes*) and raccoon dog (*Nyctereutes procyonoides*) were all consumed according to availability, both on the population ($\chi^2 = 5.98$, df = 4, P = 0.20) and the study pair level ($\chi^2 = 11.9$, df = 12, P = 0.45).

Diet selection by size

Results of food size resource use and availability are summarised in Table 3.4. The contribution of different fish sizes to eagle diets did not reflect their respective availabilities, both on the population level ($\chi^2 = 1310.4$, df = 6, P < 0.0001) and in each study pair ($\chi^2 \geq 272.9$, df = 6, P < 0.0001). Fish size selection did not significantly differ between study

Table 3.3. *Food species preferences of five study pairs of white-tailed eagles in northeastern Germany.* o_i*: proportion of species i consumed;* π_i*: proportion of species i available;* $\hat{w}_i$*: selection ratio;* ^{P}*: significantly preferred;* ^{A}*: significantly avoided;* $^*P < 0.006$*;* $^{***}P < 0.0001$ *(tested by* χ^2 *log-likelihood statistics, Bonferroni adjusted* $\alpha = 0.006$*).*

Food species	SP1 to SP3				SP4				SP5			
	o_i	π_i	$\hat{w}_i$ (SE)	χ^2	o_i	π_i	$\hat{w}_i$ (SE)	χ^2	o_i	π_i	$\hat{w}_i$ (SE)	χ^2
A) Fish												
Perch	0.019	0.175	0.11 (0.06)^{A***}	208.9	0	0.039	0	-	0.014	0.097	0.15 (0.15)^{A***}	34.0
Roach	0.132	0.588	0.22 (0.05)^{A***}	287.5	0.036	0.322	0.11 (0.06)^{A***}	194.9	0.057	0.244	0.23 (0.11)^{A***}	45.5
Rudd	0.050	0.090	0.56 (0.19)	5.16	0.036	0.080	0.45 (0.26)	4.61	0.029	0.023	1.23 (0.86)	0.07
White bream	0	0.007	0	-	0.012	0.040	0.30 (0.30)	5.28	0.029	0.035	0.82 (0.57)	0.11
Carp bream	0.660	0.123	5.37 (0.33)^{P***}	178.1	0.916	0.491	1.86 (0.06)^{P***}	182.4	0.871	0.573	1.52 (0.07)^{P***}	53.9
Pike	0.138	0.010	14.1 (3.03)^{P***}	18.8	0	0.017	0	-	0	0.015	0	-
Eel	0	0.008	0	-	0	0.011	0	-	0	0.013	0	-
No. of items (N)	159	14759			83	15417			70	10178		
B) Waterfowl												
Coot	0.418	0.074	5.64 (0.76)^{P***}	37.3	0	0	-	-	0	0	-	-
Diving duck	0.177	0.318	0.56 (0.14)^{A*}	10.8	0	0.015	0	-	0	0.018	0	-
Dabbling duck	0.051	0.059	0.87 (0.42)	0.10	0.429	0.469	0.91 (0.28)	0.09	0.200	0.106	1.88 (1.20)	0.54
Goosander	0	0.009	0	-	0.143	0.267	0.54 (0.35)	1.71	0.200	0.135	1.48 (0.94)	0.30
Grebe	0.076	0.081	0.93 (0.37)	0.03	0.429	0.166	2.58 (0.81)	3.80	0.200	0.081	2.46 (1.57)	0.87
Cormorant	0.139	0.207	0.67 (0.19)	3.04	0	0.059	0	-	0	0.082	0	-
Goose	0.127	0.221	0.57 (0.17)	6.41	0	0	-	-	0.300	0.542	0.55 (0.27)	2.79
Swan	0.013	0.030	0.42 (0.42)	1.88	0	0.025	0	-	0.100	0.035	2.87 (2.75)	0.46
No. of items (N)	79	25491			14	1503			10	1870		

Table 3.3. *continued*

Food species	SP1 to SP3				SP4				SP5			
	o_i	π_i	$\hat{w}_i$ (SE)	χ^2	o_i	π_i	$\hat{w}_i$ (SE)	χ^2	o_i	π_i	$\hat{w}_i$ (SE)	χ^2
C) Game mammals												
Fallow deer	0.167	0.215	0.77 (0.29)	0.61	0.182	0.215	0.84 (0.54)	0.08	0.167	0.215	0.77 (0.50)	0.20
Wild boar	0.472	0.347	1.36 (0.24)	2.21	0.364	0.347	1.05 (0.42)	0.01	0.417	0.347	1.20 (0.41)	0.24
Roe deer	0.278	0.352	0.79 (0.21)	0.98	0.273	0.352	0.77 (0.38)	0.35	0.167	0.352	0.47 (0.31)	2.96
Raccoon dog	0	0.021	0	-	0.091	0.021	4.23 (4.09)	0.62	0	0.021	0	-
Red fox	0.083	0.064	1.29 (0.72)	0.16	0.091	0.064	1.41 (1.35)	0.09	0.250	0.064	3.88 (1.97)	2.13
No. of items (N)	36	1815			11	1815			12	1815		

Table 3.4. *Fish size preferences of five study pairs of white-tailed eagles in northeastern Germany. o_i: proportion of fish of size i consumed; π_i: proportion of fish of size i available; $\hat{w}_i$: selection ratio. ^{P}: significantly preferred; ^{A}: significantly avoided; $^{**}P < 0.001$; $^{***}P < 0.0001$ (tested by χ^2 log-likelihood statistics, Bonferroni adjusted $\alpha = 0.007$).*

Size class (cm)	SP1 to SP3				SP4				SP5			
	o_i	π_i	$\hat{w}_i$ (SE)	χ^2	o_i	π_i	$\hat{w}_i$ (SE)	χ^2	o_i	π_i	$\hat{w}_i$ (SE)	χ^2
< 10	0	0.003	0	-	0	0.001	0	-	0	0.001	0	-
10 - 20	0.038	0.800	0.05 (0.02)^{A***}	2545.3	0.036	0.639	0.06 (0.03)^{A***}	865.5	0.057	0.678	0.08 (0.04)^{A***}	500.1
> 20 - 30	0.101	0.156	0.65 (0.15)	5.29	0.133	0.209	0.63 (0.18)	4.25	0.186	0.257	0.72 (0.18)	2.35
> 30 - 40	0.465	0.019	24.50 (2.54)^{P***}	85.8	0.602	0.112	5.40 (0.50)^{P***}	78.4	0.557	0.031	18.15 (2.18)^{P***}	61.8
> 40 - 50	0.327	0.011	31.15 (4.33)^{P***}	48.5	0.229	0.020	11.33 (2.37)^{P***}	19.0	0.200	0.014	14.60 (3.70)^{P**}	13.5
> 50 - 60	0.044	0.003	15.72 (6.30)	5.46	0	0.008	0	-	0	0.010	0	-
> 60	0.025	0.008	2.99 (1.50)	1.76	0	0.010	0	-	0	0.010	0	-
No. of items (N)	159	14759			83	15417			70	10178		

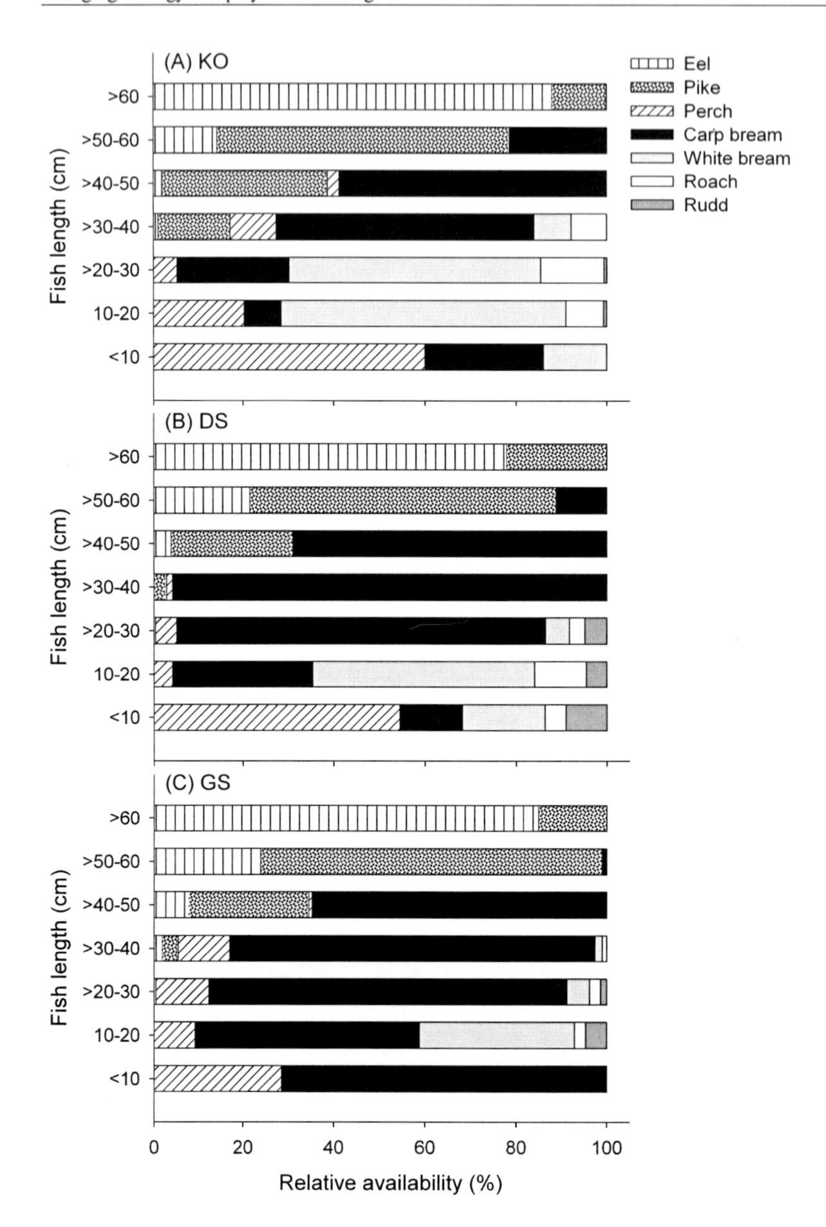

Fig. 3.3. *Relative availability of different size classes in fish species of the main water bodies used for foraging by five study pairs of white-tailed eagles in northeastern Germany. (A) Krakower Obersee (KO), (B) Damerower See (DS), (C) Goldberger See (GS).*

pairs SP1 to SP3 ($\chi^2 = 6.01$, df = 12, P = 0.92) and we therefore combined their data for the size selection analysis. All study pairs showed a similar behaviour in selecting available size classes of consumed fish species. Fish less than 10 cm were never consumed. Fish with a length of 10 to 20 cm were avoided and fish with a length of > 20 to 30 cm were consumed according to availability. Eagles preferred fish with lengths of > 30 to 40 cm and > 40 to 50 cm. The selection ratios for fish of larger size classes did not significantly differ from 1.

The different size classes were not uniformly distributed over all consumed fish species but varied significantly in each lake ($\chi^2 \geq 669.9$, df = 12, P < 0.0001). Perch and roach constituted the smallest fish species, only rarely found in size classes above > 20 to 30 cm (Fig. 3.4). Carp bream was a medium-sized species which predominantly occurred in size classes of > 30 to 40 cm and > 40 to 50 cm. Regular occurrence of pike started at the size class > 30 to 40 cm in KO, and at the size class > 40 to 50 cm in DS and GS. Eel constituted the largest fish species which dominated the size class > 60 cm.

Discussion

Time allocation and foraging strategy: Why do white-tailed eagles perch?

This is the first study examining time allocation of individual free-ranging white-tailed eagles. We found that throughout the year all study animals allocated most of their diurnal time to perching. Since our observations were restricted to the main foraging area of each focal eagle, locomotor activities such as extended foraging and extraterritorial movements documented for those study eagles wearing GPS-transmitters (Scholz 2010) were in all probability underrepresented. Inconspicuous activities such as nesting and feeding are also likely to be underestimated. Activity data of one adult white-tailed eagle fitted with a GPS-transmitter including an acceleration sensor (Krone *et al.* 2009*b*) indicated a mean daily activity of 33.1 % and thus a higher proportion of behaviours including movements than we observed. However, the extremely large proportion of perching in the eagles' time budgets and the low variance (SD = 3.3 %) among the focal eagles suggest that this behaviour dominates the diurnal activity of white-tailed eagles. Our findings are in accordance with a study on the activity pattern of bald eagles, which documented 94.3 % of perching in bald eagle time budgets (Watson *et al.* 1991), and support the observation based on spatial locations that white-tailed eagles strongly prefer habitat edges (Scholz 2010).

Why do eagles perch, or, in other words, "sit-and-wait" for their prey? Although only one-third of observed strikes resulted from perch-hunting, strike success was higher in perch-hunting than in flight-hunting. Since white-tailed eagles are large and heavy predatory birds and not nearly as agile as hawks or falcons (Rudebeck 1950), it is possible that they can secure prey by surprise only by perch-hunting, not by flight-hunting. In contrast to highly agile raptors predominantly feeding on dispersed prey such as snakes (Bakaloudis 2010), white-tailed eagles focused their foraging activity on the littoral zone of the largest water body of their home range. Consequently, sit-and-wait seems to constitute a low-cost, highly profitable foraging mode.

This result does not support the idea that larger raptors with low wing loading should use active foraging modes (Jaksić & Carothers 1985). We rarely observed prey strikes that resulted from soaring flights, considered to be the preferred foraging mode in aerial predators with relative extensive wing-span (Gamauf 1998). It is conceivable that in raptors using restricted and very open foraging areas, such as white-tailed eagles, the energetic benefits from perch-hunting exceed those from soaring flights.

Since eagles minimised energy expenditure whilst maximising energy intake, our data are consistent with the optimal foraging strategy of "energy maximising" (Schoener 1971), which is known to be adopted in a wide range of species (e.g., Karanth & Sunquist 1995; Jackson *et al.* 2004; Ichii *et al.* 2007). Minimising the time spent foraging would be a favoured strategy if factors other than foraging, such as predator avoidance, pursuit of mates and territory defence, are more important in terms of maximising fitness (Stephens & Krebs 1986; Edwards 1997). For the largest European eagle species, predation occurs only on young nestlings and is therefore of little concern to hunting eagles. Since white-tailed eagles are socially monogamous (Thiollay 1994) and all our study animals were paired and occupied a territory, they did not need to search for mates. The contrary is most likely true in non-breeders (floaters), which search for mates and have to secure a territory to produce offspring. Perching usually occurred at view points in the core area of the eagle's home range, which in our study area was relatively small with an average size of about 15 km^2 (Scholz 2010). Besides scanning for prey, eagles may also look for possible intruders during perching. We regularly observed that perched eagles detected intruders and switched to territory defence, predominantly by vocalisations and casually by chasing them. In conclusion, perching seems to be the most economic foraging strategy in territorial and breeding white-tailed eagles.

Factors influencing foraging behaviour and strike success

Foraging behaviour and strike success of white-tailed eagles was influenced both by life history features and environmental conditions. For instance, adult focal eagles exhibited a shorter foraging flight duration (FFD) and higher prey capture success and were thus more efficient hunters than juvenile eagles. This could be explained by differing knowledge in terms of local prey distribution, i.e., profitable fishing grounds, which should be much better in territorial adults than dispersing, nomadic juvenile eagles (Nygård *et al.* 2003). Consistent with this idea, all territorial eagles focused their fish capture attempts on only a few small patches at the respective lakes. Furthermore, adults probably benefit from an improvement in foraging skills with age and experience, as documented in several bird species (Edwards 1989; Wunderle 1991; Rutz *et al.* 2006).

There was also variation in strike success among lakes and study pairs – eagles foraging at KO were less successful than the ones at DS and GS. This may reflect variation in habitat quality. KO was considerable deeper and more eutrophic than the polytrophic shallow water bodies DS and GS (Table 3.1) and hosted substantially larger accumulations of waterfowl species than DS and GS. Consequently, study pairs SP1 to SP3 at KO included a distinctively larger proportion of waterfowl in their diet than SP4 and SP5 (M. Nadjafzadeh *et al.* unpublished data), which are more difficult to capture than fish (Fig. 3.2).

Our results highlight seasonal changes as important environmental factors affecting the foraging pattern of white-tailed eagles. Eagle FFD was lowest in spring and increased significantly over summer and autumn towards winter months. The spawning of fish species consumed by the eagles occurs in spring and leads to an abundant fish supply in the littoral zone (Bauch 1966) that might account for the minimised foraging duration of eagles in spring. During periods with high energy demands such as wintering, when prey availability decreases and thermoregulatory requirements increase, time and energy allocated to foraging is expected to increase (Lundberg 1985). However, considering the drastically reduced strike success in winter despite the observed increase in FFD, it is unclear as to how eagles secure their energy requirements during this season. A study conducted in the same study area and time period and with the same study animals showed that all study animals exhibited a functional response to the decline of fish and waterfowl availability in winter and switched their diet to alternative food sources such as carcasses of game mammals (M. Nadjafzadeh *et al.* unpublished data). As these food sources are unpredictably distributed, they require widely foraging (Huey & Pianka 1981). This suggests that white-tailed eagles can modify their foraging strategy and shift from sit-and-wait hunting towards an active search of scavenging items to cope with variations of weather conditions and food availability. Such flexibility in foraging behaviour has also been documented in large carnivores which pursue energy maximising strategies in prey-rich environments and are forced to increase foraging effort in prey-poor environments to ensure survival (Karanth & Sunquist 1995; Bothma & Coertze 2004).

Dietary preferences

The present study demonstrates that white-tailed eagles, previously categorised as catholic predators (Oehme 1975; Helander 1983; Sulkava *et al.* 1997), hunted non-randomly and focused their foraging on particular prey species. Within their primary prey, fish, all study pairs showed a strong preference for carp bream and avoidance of perch and roach. This strong selection of species may be explained by the distribution of size classes across the different species. Whereas perch and roach body sizes were predominantly within the small size classes of < 10 to 20 cm, avoided by all study pairs, carp bream comprised the majority of fish available in large size classes of > 30 to 50 cm, which eagles strongly preferred. Pike, which occurred in the KO regularly within the preferred size classes, was also significantly preferred by the resident study pairs. This outcome is consistent with the prediction that large predators should prefer larger-bodied prey (Pyke *et al.* 1977). Fish larger than 50 cm were not consumed (SP4, SP5), or consumed according to availability (SP1 to SP3). The optimal diet model predicts that the degree of selection depends on the encounter rate with profitable (large) prey (Krebs 1978; Barnard & Brown 1981). Since the absolute availability of fish larger than 50 cm was low, it would probably not benefit eagles to specialise on this size class. Several prey size selection studies showed that the largest prey a predator can handle is not necessarily the most profitable one, but prey of intermediate to larger size (Karanth & Sunquist 1995; Mikheev & Wanzenböck 1999; Gil & Pleguezuelos 2001).

White-tailed eagles also discriminated amongst species in alternative prey classes such as waterfowl, provided species were abundant as at the KO. The KO study pairs took coots significantly more frequently and diving ducks significantly less frequently than expected. Coots represented the smallest (36-42 cm) and diving ducks the second smallest (40-47 cm) available waterfowl species to contribute to the eagle diet (Svensson *et al.* 1999). Hence, contrary to fish, size seems not to be the crucial factor in waterfowl selection. Since coots as well as diving ducks live in flocks, detectability (Ioannou & Krause 2008) and vigilance

(Beauchamp 2003) should be similar for both species and not influence eagle prey choice. Coots largely differ from diving ducks by weak diving and flying abilities (Struwe-Juhl 2003). This probably renders coots a prey easier to catch. The opportunistic foraging behaviour for game carrion observed in all study pairs makes sense because of its extremely ephemeral nature and spatial and temporal patchiness (DeVault *et al.* 2003).

In conclusion, we demonstrated in this study that despite their high dietary diversity, white-tailed eagles are selective predators that harvest their prey in accordance with predictions of optimal foraging theory in terms of absolute availability, size and anti-predator behaviour.

Implications for conservation

The key role of perches in the foraging behaviour of white-tailed eagles may constitute a disadvantage for their survival and dispersion in highly altered landscapes such as most rural Germany. Since our study clearly indicates that riparian trees are a key element of the hunting modes of white-tailed eagles, these foraging "tools" should be preserved in the context of landscaping and forestry use. If logging is inevitable in areas inhabited by white-tailed eagles, our findings suggest that the conservation of riparian wooded corridors of around 100 m in width would be mandatory to protect the most relevant perches and thus the core zone within the hunting grounds of eagles.

The diet selection analyses revealed that white-tailed eagles prefer fish species of reduced interest to fishermen, such as carp bream, and avoid European perch or eel which represent important species to fisheries. Thus, increasing populations of white-tailed eagles at lakes used for commercial fishery should not lead to a conflict of interest as in the case of great cormorants (Frederiksen *et al.* 2001).

Waterfowl species such as Eurasian coot can constitute a major and highly preferred food resource for white-tailed eagles. Although this species does not face conservation problems in general, local populations – including the ones in our study area – have drastically decreased during the last two decades (Neubauer 2001). As white-tailed eagles appear flexible in their diet selection provided alternative prey species are available, they will probably not be strongly affected by changing abundances of single species. However, successful foraging is a serious challenge for white-tailed eagles in periods with low prey availability such as winter months. Opportunistic consumption of game carrion by eagles during such times appears to be their method to cope with scarcity of their usual prey. Unfortunately, game carrion actually poses a severe risk to white-tailed eagles as it is the main source of lead bullet fragments that induce fatal lead poisoning (Hunt *et al.* 2006; Krone *et al.* 2009*a*; M. Nadjafzadeh *et al.* unpublished data).

Acknowledgements

We are grateful to the administration of the nature park *Nossentiner/Schwinzer Heide* and the *Reepsholt-Stiftung* for logistic support and accommodation. We thank W. Neubauer for data on waterfowl availability at the *Krakower Obersee*. We are indebted to several fishermen, forestry districts, hunters and landowners in the study area for collaboration and support, and to F. Scholz, J. Sulawa, A. Trinogga, and N. Kenntner for assistance and support. This study was funded by the Federal Ministry of Education and Research (BMBF, reference no. 0330720) and the Leibniz Institute for Zoo and Wildlife Research Berlin.

References

Bakaloudis, D.E. 2010. Hunting strategies and foraging performance of the short-toed eagle in the Dadia-Lefkimi-Soufli National Park, north-east Greece. *Journal of Zoology* 281: 168-174.

Barnard, C.J. & Brown, C.A.J. 1981. Prey size selection and competition in the common shrew (*Sorex araneus* L.). *Behavioral Ecology and Sociobiology* 8: 239-243.

Beauchamp, G. 2003. Group-size effects on vigilance: a search for mechanisms. *Behavioural Processes* 63: 111-121.

Bibby, C.J. 1995. *Methoden der Feldornithologie*. Neumann Verlag, Radebeul, Germany.

Bothma, J. du P. & Coertze, R.J. 2004. Motherhood increases hunting success in southern Kalahari leopards. *Journal of Mammalogy* 85: 756-760.

DeVault, T.L., Rhodes, O.E. & Shivik, J.A. 2003. Scavenging by vertebrates: behavioral, ecological, and evolutionary perspectives on an important energy transfer pathway in terrestrial ecosystems. *Oikos* 102: 224-234.

Edwards, G.P. 1997. Predicting seasonal diet in the yellow-bellied marmot: success and failure for the linear programming model. *Oecologia* 112: 320-330.

Edwards, T.C. 1989. The ontogeny of diet selection in fledgling ospreys. *Ecology* 70: 881-896.

Fahmy, E.-G. 1982. Untersuchungen zur Bestandscharakteristik und Populationsdynamik des Bleis (*Abramis brama* L.) sowie seiner Einordnung in das Trophiegefüge der Darß-Zingster-Boddenkette. *Doctoral dissertation*, Universität Rostock, Rostock, Germany.

Fischer, W. 1982. *Die Seeadler*. Ziemsen Verlag, Wittenberg Lutherstadt, Germany.

Frederiksen, M., Lebreton J.-D. & Bregnballe, T. 2001. The interplay between culling and density-dependence in the great cormorant: a modelling approach. *Journal of Applied Ecology* 38: 617-627.

Gamauf, A., Preleuthner, M. & Winkler, H. 1998. Philippine birds of prey: interrelations among habitat, morphology and behavior. *Auk* 75: 713-726.

Gil, J.M. & Pleguezuelos, J.M. 2001. Prey and prey-size selection by the short-toed eagle (*Circaetus gallicus*) during the breeding season in Granada (south-eastern Spain). *Journal of Zoology* 255: 131-137.

Goss-Custard, J.D. 1977. Optimal foraging and the size selection of worms by redshank, *Tringa totanus*, in the field. *Animal Behaviour* 25: 10-29.

Hauff, P. 2003. Sea-eagles in Germany and their population growth in the 20th century. In: *Sea eagle 2000*, (eds.) Helander, B., Marquiss, M. & Bowerman, B., pp. 71-77. Swedish Society for Nature Conservation/SNF, Stockholm, Sweden.

Helander, B. 1983. Reproduction of the white-tailed sea eagle *Haliaeetus albicilla* (L.) in Sweden, in relation to food and residue levels of organochlorine and mercury compounds in the eggs. *Ph.D. dissertation*, University of Stockholm, Stockholm, Sweden.

Helander, B. & Sternberg, T. 2002. *Action plan for the conservation of white-tailed sea eagles* (*Haliaeetus albicilla*). Birdlife International, Strasbourg, France.

Hixon, M.A. & Carpenter F.L. 1988. Distinguishing energy maximizers from time minimizers: a comparative study of two hummingbird species. *Integrative and Comparative Biology* 28: 913-925.

Huey, R.B. & Pianka, E.R. 1981. Ecological consequences of foraging mode. *Ecology* 62: 991-999.

Hunt, W.G., Burnham, W., Parish, C.N., Burnham, B., Mutch, B. & Oaks, J.L. 2006. Bullet fragments in deer remains: implications for lead exposure in scavengers. *Wildlife Society Bulletin* 34: 167-170.

Ichii, T., Bengtson, J.L., Boveng, P.L., Takao, Y., Jansen, J.K., Hiruki-Raring, L.M., Cameron, M.F., Okamura, H., Hayashi, T. & Naganobu, M. 2007. Provisioning strategies of Antarctic fur seals and chinstrap penguins produce different responses to distribution of common prey and habitat. *Marine Ecology Progress Series* 344: 277-297.

Ioannou, C.C. & Krause, J. 2008. Searching for prey: the effects of group size and number. *Animal Behaviour* 75: 1383-1388.

Jackson, A.C., Rundle, S.D., Attrill, M.J. & Cotton, P.A. 2004. Ontogenetic changes in metabolism may determine diet shifts for a sit-and-wait predator. *Journal of Animal Ecology* 73: 536-545.

Jaksić, F.M. & Carothers, J.H. 1985. Ecological, morphological, and bioenergetic correlates of hunting mode in hawks and owls. *Ornis Scandinavica* 16: 165-172.

Jenkins, A.R. 2000. Hunting mode and success of African peregrines *Falco peregrinus minor*: does nesting habitat quality affect foraging efficiency? *Ibis* 142: 235-246.

Karanth, K.N. & Sunquist, M.E. 1995. Prey selection by tiger, leopard and dhole in tropical forests. *Journal of Animal Ecology* 64: 439-450.

Krebs, J.R. 1978. Optimal foraging: decision rules for predators. In: *Behavioral ecology: an evolutionary approach*, (eds.) Krebs, J.R. & Davies, N.B., pp. 23-63. Blackwell Scientific Publications, Oxford, UK.

Krone, O., Kenntner, N. & Tataruch, F. 2009*a*. Gefährdungsursachen des Seeadlers (*Haliaeetus albicilla* L. 1758). *Denisia* 27: 139-146.

Krone, O., Berger, A. & Schulte, R. 2009*b*. Recording movement and activity pattern of a white-tailed sea eagle (*Haliaeetus albicilla*) by a GPS datalogger. *Journal of Ornithology* 150: 273-280.

Lemon, W.C. 1991. Fitness consequences of foraging behaviour in the zebra finch. *Nature* 352: 153-155.

Lundberg, P. 1985. Time-budgeting by starlings *Sturus vulgaris*: Time minimising, energy maximising and the annual cycle organization. *Journal of Animal Ecology* 67: 331-337.

Manly, B.F.J., McDonald, L.L., Thomas, D.L. & McDonald, T.L. 2002. *Resource selection by animals*. 2nd edition. Kluwer Academic Publishers, Dordrecht, Netherlands.

Martin, P. & Bateson, P. 1993. *Measuring behaviour*. 2nd edition. Cambridge University Press, Cambridge, UK.

Masman, D., Daan, S. & Dijkstra, C. 1988. Time allocation in the kestrel (*Falco tinnunculus*), and the principle of energy minimization. *Journal of Animal Ecology* 57: 411-432.

Mikheev, V.N. & Wanzenböck, J. 1999. Satiation-dependent, intra-cohort variations in prey size selection of young roach (*Rutilus rutilus*). *Oecologia* 121: 499-505.

Naef-Daenzer, L., Naef-Daenzer, B. & Nager, R.G. 2000. Prey selection and foraging performance of breeding great tits *Parus major* in relation to food availability. *Journal of Avian Biology* 31: 206-214.

Neubauer, W. 2001. *Die Vögel des Naturschutzgebietes Krakower Obersee*. Oemke Verlag, Gützkow, Germany.

Nygård, T., Kenward, R.E. & Einvik, K. 2003. Dispersal in juvenile white-tailed sea eagles in Norway shown by radio-telemetry. In: *Sea eagle 2000*, (eds.) Helander, B., Marquiss, M. & Bowerman, B., pp. 191-196. Swedish Society for Nature Conservation/SNF, Stockholm, Sweden.

Oehme, G. 1975. Ernährungsökologie des Seeadlers, *Haliaeetus albicilla* (L.), unter besonderer Berücksichtigung der Population in den drei Nordbezirken der Deutschen Demokratischen Republik. *Doctoral dissertation*, Universität Greifswald, Greifswald, Germany.

Oksanen, L. & Lundberg, P. 1995. Optimization of reproductive effort and foraging time in mammals: the influence of resource level and predation risk. *Evolutionary Ecology* 9: 45-56.

Perry, G. 1999. The evolution of search modes: Ecological versus phylogenetic perspectives. *American Naturalist* 153: 98-109.

Pyke, G.H., Pulliam, H.R. & Charnov, E.L. 1977. Optimal foraging: a selective review of theory and tests. *The Quarterly Review of Biology* 52: 137-154.

Rudebeck, G. 1950. The choice of prey and modes of hunting of predatory birds with special reference to their selective effect. *Oikos* 2: 65-88.

Rutz, C., Wittingham, M.J. & Newton, I. 2006. Age-dependent diet choice in an avian top predator. *Proceedings of the Royal Society B* 273: 579-586.

Schoener, T.W. 1971. Theory of feeding strategies. *Annual Review of Ecology and Systematics* 2: 369-404.

Scholz, F. 2010. Spatial use and habitat selection of white-tailed eagles (*Haliaeetus albicilla*) in Germany. *Doctoral dissertation*, Freie Universität Berlin, Berlin, Germany.

Shepard, E.L.C., Wilson, R.P., Quintana, F., Laich, A.G. & Forman, D.W. 2009. Pushed for time or saving on fuel: fine-scale energy budgets shed light on currencies in a diving bird. *Proceedings of the Royal Society B* 276: 3149-3155.

Sinclair, A.R.E. & Krebs, C.J. 2002. Complex numerical responses to top-down and bottom-up processes in vertebrate populations. *Philosophical Transactions of the Royal Society B* 357: 1221-1231.

Stephens, D.W. & Krebs, J.R. 1986. *Foraging theory*. Princeton University Press, Princeton, USA.

Struwe-Juhl, B. 2003. Why do white-tailed eagles prefer coots? In: *Sea eagle 2000*, (eds.) Helander, B., Marquiss, M. & Bowerman, B., pp. 317-326. Swedish Society for Nature Conservation/SNF, Stockholm, Sweden.

Sulkava, S., Tornberg, R. & Koivusaari, J. 1997. Diet of the white-tailed eagle *Haliaeetus albicilla* in Finland. *Ornis Fennica* 74: 65-78.

Svensson, L., Grant, P.J., Mullarney, K. & Zetterström, D. 1999. *Der neue Kosmos-Vogelführer. Alle Arten Europas, Nordafrikas und Vorderasiens*. Franckh-Kosmos Verlags-GmbH & Co. KG, Stuttgart, Germany.

Tarkan, A.S., Gürsoy Gaygusuz, Ç., Gaygusuz, Ö. & Acipinar, H. 2007. Use of bone and otolith measures for size-estimation of fish in predator-prey studies. *Folia Zoologica* 56: 328-336.

Teerink, B.J. 1991. *Hair of west European mammals*. Cambridge University Press, Cambridge, UK.

Thiollay, J.M. 1994. Family Accipitridae (hawks and eagles). In: *Handbook of the birds of the world: New World vultures to guineafowl*. Volume 2, (eds.) del Hoyo, J., Elliot, A. & Sargatal, J., pp. 52-205. Lynx Edicions, Barcelona, Spain.

Thomas, D.L. & Taylor, E.J. 1990. Study designs and tests for comparing resource use and availability. *Journal of Wildlife Management* 54: 322-330.

van Gils, J.A., Dekinga, A., Spaans, B., Vahl, W.K. & Piersma, T. 2005. Digestive bottleneck affects foraging decisions in red knots *Calidris canutus*. II. Patch choice and length of working day. *Journal of Animal Ecology* 74: 120-130.

Watson, J.W., Garrett, M.G. & Anthony, R.G. 1991. Foraging ecology of bald eagles in the Columbia River estuary. *Journal of Wildlife Management* 55: 492-499.

Wunderle, J.M. 1991. Age-specific foraging proficiency in birds. *Current Ornithology* 8: 273-324.

CHAPTER 4

Specialisation in a generalist predator? Evidence from stable isotope analysis for spatial, seasonal and individual variation in diet composition of white-tailed eagles

Abstract

Raptors are usually regarded as generalists when consuming a wide array of prey, although the required evidence, longitudinal dietary records of populations and individuals, are usually not available. A comprehensive quantification of niche width in top-end predators such as white-tailed eagles (*Haliaeetus albicilla*) should include food types like carrion and is difficult by conventional means but often essential for refining conservation efforts. Many raptors, and white-tailed eagles in particular, are affected by lead poisoning probably caused by ingestion of game mammal carcasses containing fragments of lead-based ammunition. Therefore, a robust quantitative assessment of the relevance of carrion in their diet is urgently needed. We used stable isotopes ($\delta^{13}C$, $\delta^{15}N$) for detecting patterns of dietary variation and quantifying diet composition in the white-tailed eagle. We analysed the isotopic composition of tissues from eagles of the German population ($n = 75$) and of 16 potential food species ($n = 90$), and the Finnish ($n = 10$) and Greenlandic ($n = 10$) populations. Spatial comparison revealed significant isotopic differences between the German and Greenlandic populations and reflected the main foraging areas of each population. Within the German population, we found significant (1) seasonal isotopic differences, suggesting dietary responses caused by changing food availabilities, and (2) age-related isotopic differences indicating altered resource-use efficiencies in adults and juveniles. Isotopic signatures of liver and muscle tissues showed significant intra-individual short-term changes in the German and Finnish but not Greenlandic population. This suggests that local feeding niches of white-tailed eagles vary with local food supplies and indicates both individual generalisation caused by dietary shifts and specialisation because of constant foraging patterns. Using mass-balance (IsoError) and Bayesian (SIAR) mixing models, we infer that game mammal carcasses constitute important alternative food sources for the German eagle population during the hunting season. Thus, our findings clearly demonstrate that lead-contaminated carcasses can pose a serious hazard to white-tailed eagles.

Keywords: Bayesian mixing models, carbon isotopes, feeding habits, *Haliaeetus albicilla*, IsoError, isotopic niche, nitrogen isotopes, SIAR

Introduction

Feeding habits are a central component of the biology of a species, because of their relevance to survival, reproduction and population dynamics. Knowledge about the diet composition of animals is of particular importance for their effective conservation management (Sinclair *et al.* 2006). Common methods applied in dietary studies mostly consist of the analysis of prey remains, pellet or stomach contents. These methodologies can provide detailed information on individual food consumption but in general reflect short-term intake (Votier *et al.* 2003). During the last decades, the use of stable isotope analyses has been increasingly adopted as an alternative approach to draw quantitative conclusions about the relative contributions of potential food sources to a species' diet (Hobson 1999; Inger & Bearhop 2008). Typically, stable isotope ratios in animal tissue are related to those of their diet (DeNiro & Epstein 1978). The ratios of naturally occurring isotopes of carbon (^{13}C and ^{12}C) and nitrogen (^{15}N and ^{14}N) can vary between marine and freshwater food webs, aquatic and terrestrial habitats, and different trophic levels (DeNiro & Epstein 1981; Angerbjörn *et al.* 1994; Hobson 1999; McCutchan *et al.* 2003). Recent developments in stable isotope mixing models (SIAR, Parnell *et al.* 2010) allow robust estimations for the range of contributions of multiple food sources to a consumer's diet. Furthermore, different types of animal tissues vary in the rate at which they incorporate new materials (Tieszen *et al.* 1983). Consequently, the comparison among the isotopic composition of tissues with dissimilar incorporation rates of a single individual can provide information about its degree of dietary specialisation (Martínez del Rio *et al.* 2009).

The use of stable isotope analysis is particularly useful for obtaining information from organisms that are difficult to observe (Cherel *et al.* 2000) such as raptor species which regularly forage over large and remote areas (Austin *et al.* 1996). Although conventional techniques for analysing raptor diets are biased in various ways because of overrepresentation or underrepresentation of particular food types (Mersmann *et al.* 1992), stable isotope analyses are currently rarely applied (but see Chamberlain *et al.* 2005; Caut *et al.* 2006; Newsome *et al.* 2010). In case of the white-tailed eagle (*Haliaeetus albicilla*), several conventional dietary studies exist but with highly varying results in terms of their main food types fish, waterfowl, and mammalian carrion (Oehme 1975; Helander 1983; Wille & Kampp 1983; Sulkava *et al.* 1997; Struwe-Juhl 2003). Since different methods were used, it is difficult to evaluate to what extent such differences may reflect spatial, seasonal or individual variation.

Currently, data on individual feeding habits are not available for white-tailed eagles. Owing to their diverse array of prey species, white-tailed eagles are considered to be dietary generalists. However, a large niche width can be the result of coexisting individual generalists and specialists, or of individual specialists differing in their diet composition over a larger geographical area (Bolnick *et al.* 2003). Based on current knowledge, we predicted that the white-tailed eagle should exhibit a broad range of isotope values, representing a large niche width and flexible foraging tactics.

Quantitative measurements of the diet composition of scavenging raptors provide crucial information for optimising conservation efforts with regard to lead poisoning. This anthropogenic threat affects numerous species (Fisher *et al.* 2006) and is the major cause of death in white-tailed eagles (Krone *et al.* 2006; Krone *et al.* 2009; Helander *et al.* 2009). It is assumed that carcasses and gut piles from hunter-killed game mammals, containing fragments of lead-based ammunition and often left in the environment, are the main sources of lead for avian scavengers (Pain & Amiard-Triquet 1993; Hunt *et al.* 2006). However, a realistic estimation of the contribution of carrion to the diet of raptors is difficult to achieve with conventional techniques, since these food sources occur at unpredictable locations and the ingestion of pure flesh such as gut piles cannot be traced by remains in pellets. Stable isotope analysis can provide reliable information about the consumption of game mammal carrion in white-tailed eagles, since game mammals represent terrestrial food sources and should therefore be isotopically distinct from the usual prey, fish and waterfowl, from aquatic habitats.

In this study we provide quantitative data on the feeding habits of white-tailed eagles using stable isotope analysis in order to answer the following questions: (1) does the diet composition of white-tailed eagles vary between different populations, individuals or seasons, (2) to what extent are white-tailed eagles dietary generalists, and (3) how important is terrestrial food such as carrion in the diet of white-tailed eagles? We then used the answers to these questions to assess the conservation implications of eagle feeding habits.

Material and methods

Study areas

We studied white-tailed eagle populations in Germany, Finland, and Greenland (Fig. 4.1). Presently, the German eagle population is confined to the northeast of Germany and consists of approximately 570 breeding pairs (Hauff 2008). Here, the federal states of Mecklenburg-Western Pomerania and Brandenburg contain about 70 % of the total population. The region is characterised by a well-preserved natural environment with numerous lakes, rivers and a high proportion of planted pinewood and mixed forest. Finnish eagles inhabit both Lapland in northern Finland and the Baltic coast in southern Finland. Their population size was estimated at 200 breeding pairs in 2000 (Stjernberg *et al.* 2003). In the archipelago of southwestern Finland, the numerous islands are relatively high and rocky and the surrounding waters relatively deep (Sulkava *et al.* 1997). Eagle pairs from the Baltic fringe predominantly nest in the vicinity of brackish waters. In contrast, Lapland constitutes a comparatively flat area and is characterised by numerous smaller freshwater lakes and large open bogs. Both areas are dominated by pine forest. The eagle population in Greenland is confined to the southwest coast and was estimated at 160 breeding pairs in 2000 (Wille 2003). The massive ice sheet covering about 80 % of the island's surface prevents the occupation of inland habitats by eagles.

Eagle sample collection

For the analysis of diet composition and possible dietary specialisation of white-tailed eagles in different geographic areas, we examined liver and muscle tissues of individuals found dead in Germany, Finland and Greenland between 1996 and 2003 (Fig. 4.1). Liver and muscle tissue reflect the isotopic composition of resources incorporated over a few days to two weeks and from two weeks to one month, respectively (Martínez del Rio *et al.* 2009). In order to compare isotopic signatures of individuals of the same country, differentiate between seasons and estimate the contributions of multiple sources via stable isotope mixing models, we continued the collection of muscle tissues of German eagles until 2008. Eagle carcasses were sent for necropsies to the Leibniz Institute for Zoo and Wildlife Research (IZW) in Berlin and stored at -20°C. In case of Finnish and Greenlandic eagles, the Finnish Museum of Natural History and the Greenland Nature Institute dispatched them to the IZW by air cargo so that all carcasses arrived in frozen condition. We included in our sample only tissues from individuals that were in good nutritional state to prevent any bias towards diseased eagles which might have been restricted in terms of foraging or capturing prey.

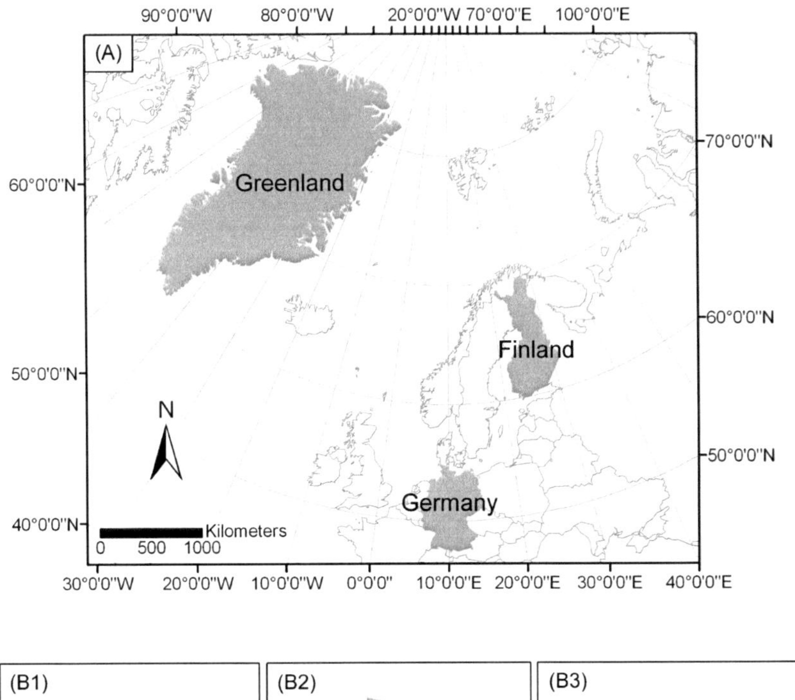

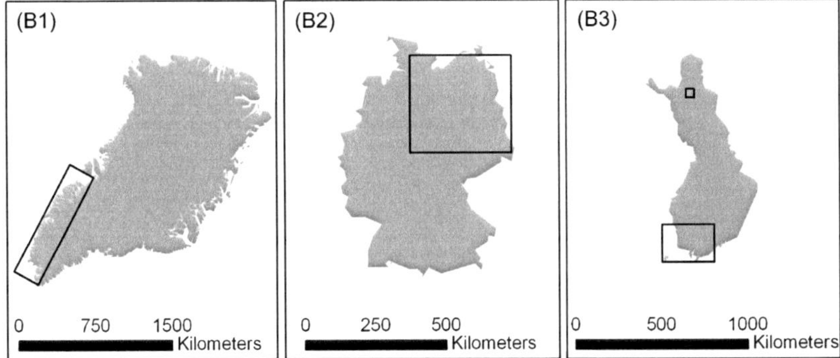

Fig. 4.1. *(A) Location of the study countries Greenland, Germany and Finland. (B) Location of the sample collection areas (black framed boxes) in southwestern Greenland (B1), northeastern Germany (B2), and southern and northern Finland (B3).*

Food source sample collection

We used a priori knowledge of the diet and food availability of white-tailed eagles in northeastern Germany obtained in a detailed analysis of food remains, pellet and stomach contents and simultaneous monitoring of prey populations (M. Nadjafzadeh *et al.* unpublished data) to select the most likely food sources. These comprised five fish and six waterfowl species, which would be captured alive by white-tailed eagles, and five game mammal species representing ingested carrion (Table 4.1). We collected muscle tissue samples in the core distribution area of German eagles, i.e. Mecklenburg-Western Pomerania. Muscle samples of fish species were provided by fishermen and samples of waterfowl and game mammal species originated from hunters of the study area. Additional waterfowl muscle samples were provided by the Berlin Museum of Natural History.

Stable isotope analysis

All tissues were kept frozen until preparation for determination of stable isotope ratios. Prior to analysis, samples were freeze-dried and then ground to a homogenous powder with a ceramic mortar. Subsequently, 0.5 mg sub-samples were weighed into tin cups and combusted in a Flash Elemental Analyser (Thermo Finnigan, Bremen, Germany). Further analyses of resultant N_2 and CO_2 gases were performed using a Delta V Advantage isotope ratio mass spectrometer (Thermo Fisher, Bremen, Germany). The isotope ratios were reported as δ-values and expressed as relative difference per mil (‰) according to the equation

$$\delta X = (R_{sample} / R_{standard} - 1) \times 1000$$

where X is ^{13}C or ^{15}N, R_{sample} is the corresponding ratio $^{13}C/^{12}C$ or $^{15}N/^{14}N$, and $R_{standard}$ is the ratio of the international references Pee Dee Belemnite (V-PDB) for carbon and atmospheric N_2 (AIR) for nitrogen. Analytical error was ± 0.1 ‰ for $\delta^{13}C$ and $\delta^{15}N$ values.

Mixing models

To provide a detailed examination of the food composition of the German white-tailed eagle population, we investigated their diet on three different levels. This involved the combination of potential food sources a priori, since the lumping of sources allows inferences about the dietary importance of a logically defined aggregate source (Phillips *et al.* 2005). Firstly, to determine the amount of terrestrial prey in the diet of white-tailed eagles, we combined the potential food sources based on their main habitat resulting in the following two sources: species of aquatic habitats and species of terrestrial habitats (Table 4.1). Secondly, to determine the relative contribution of the three major food sources to the diet of white-tailed eagles, we combined the potential food sources based on their taxonomic classification into the following three sources: fish, waterfowl and game mammals (Table 4.1). In order to prevent biased results due to an emphasis on different trophic levels, we averaged in each taxonomically aggregated source over herbivorous, omnivorous and piscivorous/carnivorous species. Finally, to estimate the importance of particular species in the diet of white-tailed eagles, we selected five very likely food species with significantly distinct isotopic signatures (Mann-Whitney U-test, $P_{exact} \leq 0.04$): The fish species pike (*Esox lucius*) and roach (*Rutilus rutilus*), the waterfowl species mallard duck (*Anas platyrhynchos*) and coot (*Fulica atra*), and as game mammal species wild boar (*Sus scrofa*). We analysed the diet composition for each level on a seasonal basis by pooling the samples over the respective months. Seasons were defined as winter half-year (September-February) and summer half-year (March-August).

We used dietary mixing models as a tool to determine the isotopic contribution of the food sources in the eagle tissues and thus determine the relative contributions to the eagle diet. For n isotopes, a unique mathematical solution can be calculated for up to $n + 1$ sources. Accordingly, we adopted a two source linear mixing model to examine the partition of aquatic and terrestrial food sources, using the isotopic signature of $\delta^{15}N$, and a three source linear mixing model to determine the partition of the major food sources fish, waterfowl and game mammals, using the isotopic signature of $\delta^{13}C$ and $\delta^{15}N$. We applied the IsoError procedure described by Phillips and Gregg (2001) in order to calculate variances, standard errors (SE), and 95 % confidence intervals (CI) for source proportion estimates that account for the variability in the isotopic signatures of the food sources as well as the eagles.

To estimate the relative contribution of multiple sources to the eagle diet, we employed a novel Bayesian isotope mixing model, SIAR (SIAR: Stable Isotope Analysis in R, Parnell *et al.* 2010), since it provides advantages over standard, mass-balance multi-source mixing models (e.g. IsoSource, Phillips & Gregg 2003): it can integrate sources of variability associated with multiple sources, fractionation or isotopic signatures, and its outputs represent true probability density functions, rather than a range of feasible solutions. In contrast to MixSIR, another recently published Bayesian isotope mixing model (Moore & Semmens 2008), SIAR includes an overall residual error term which accounts for unknown sources of error on the observed data.

Table 4.1. *Mean $\delta^{13}C$ and $\delta^{15}N$ muscle values, habitat use and trophic status of 16 potential food species for the German white-tailed eagle population collected in the federal state of Mecklenburg-Western Pomerania in northeastern Germany.*

Common name	Species	Abbreviation in Figure	n	Aquatic (A)/ Terrestrial (T)*	Trophic status*†	$\delta^{13}C$ [‰] (± SD)	$\delta^{15}N$ [‰] (± SD)
Fish							
European perch	*Perca fluviatilis*	1	5	A	P	-26.8 (± 0.5)	14.4 (± 1.7)
Northern pike	*Esox lucius*	2	6	A	P	-25.2 (± 0.6)	15.1 (± 0.4)
Carp bream	*Abramis brama*	3	6	A	O	-27.7 (± 0.5)	11.8 (± 2.4)
Common roach	*Rutilus rutilus*	4	6	A	O	-26.1 (± 0.6)	13.7 (± 0.6)
Common rudd	*Scardinius erythrophthalmus*	5	6	A	O/H	-24.2 (± 1.9)	12.3 (± 1.2)
Waterfowl							
Great cormorant	*Phalacrocorax carbo*	6	6	A	P	-26.2 (± 3.1)	15.8 (± 1.6)
Great crested grebe	*Podiceps cristatus*	7	1	A	P	-22.9	16.2
Black-headed gull	*Larus ridibundus*	8	6	A/T	P/O	-24.3 (± 0.8)	10.7 (± 0.5)
Mallard duck	*Anas platyrhynchos*	9	5	A/T	O	-27.0 (± 1.6)	9.9 (± 1.5)
Eurasian coot	*Fulica atra*	10	6	A/T	H/O	-30.1 (± 1.5)	11.3 (± 3.4)
Goose	*Anser* spec.	11	7	A/T	H	-28.5 (± 1.4)	7.6 (± 0.7)
Game mammals							
Red fox	*Vulpes vulpes*	12	6	T	C/O	-24.6 (± 0.7)	9.0 (± 2.3)
Raccoon dog	*Nyctereutes procyonoides*	13	6	T	C/O	-24.4 (± 0.7)	10.6 (± 1.9)
Wild boar	*Sus scrofa*	14	6	T	O	-23.1 (± 2.1)	1.9 (± 2.1)
Fallow deer	*Cervus dama*	15	6	T	H	-27.4 (± 0.4)	2.5 (± 0.9)
Roe deer	*Capreolus capreolus*	16	6	T	H	-27.4 (± 0.7)	-0.5 (± 1.3)

*First letter indicates predominant used habitat and trophic status where two are present.

†P: piscivore; C: carnivore; O: omnivore; H: herbivore.

To correct for the enrichment in predator's stable isotope ratios compared with its diet, we used discrimination factors (DeNiro & Epstein 1981; Tieszen & Botton 1988) and added them to the isotopic signatures of all food sources before integrating into IsoError or included them separately into SIAR. In accordance with the findings of Caut *et al.* (2009), we accounted for taxon, diet and tissue type by estimating discrimination values from literature sources, and determined a mean value of 1.1 ± 0.9 ‰ for $\delta^{13}C$ and 2.0 ± 0.5 ‰ for $\delta^{15}N$ based on published studies using muscle samples of piscivorous/carnivorous captive bird species (Caut *et al.* 2009).

Statistical analysis

To assess differences in $\delta^{13}C$ and $\delta^{15}N$ among German, Finnish and Greenlandic white-tailed eagle populations, we first performed a multivariate analysis of variance (MANOVA) and used Wilk's lambda to evaluate the results. If any significant difference was detected, we applied subsequently a one-way analysis of variance (ANOVA) to make overall comparisons of individual variables among populations. Tukey post hoc tests were used for pair-wise comparisons of samples. To test whether the isotopic niche of the German, Finnish or Greenlandic eagle population changed within a time frame of several weeks, we compared the isotopic composition among liver and muscle tissues for each population. We first used repeated measures MANOVA to test for differences in $\delta^{13}C$ and $\delta^{15}N$ among tissues. These analyses were followed by repeated measures ANOVA on $\delta^{13}C$ and $\delta^{15}N$, respectively. To determine factors influencing isotope ratios within the German white-tailed eagle population, we used a three-way ANOVA to test for the effects of season (winter half-year, summer half-year), sex, and age (juvenile, adult) and, because of small sample sizes (< 10), applied a non-parametric analysis of variance (Kruskal-Wallis test) to test the effects of federal state and year. To evaluate differences in isotope ratios among food sources, we used Mann-Whitney U-tests. We tested whether data deviated from normality using Kolmogorov-Smirnov tests, and whether data deviated from homoscedasticity using Levene's tests. Statistical analyses were carried out in SPSS version 16.0 (SPSS Inc., Chicago, IL). All tests were two-tailed with accepted significance levels of $P < 0.05$. Results are reported as means ± 1 standard deviation (SD) unless otherwise noted.

Results

Comparison between populations

For the period between 1996 and 2003, liver and muscle tissues of 30 white-tailed eagles from Germany, 10 eagles from Finland and 10 eagles from Greenland were available, resulting in 100 data points each for $\delta^{13}C$ and $\delta^{15}N$. Overall, mean isotopic signatures differed significantly between these geographically separated populations for combined isotopes (MANOVA: Wilks' λ = 0.84, $F_{2,99}$ = 4.36, P = 0.002) and separately for $\delta^{15}N$ isotopes (ANOVA: $F_{1,99}$ = 4.93, P = 0.009; Fig. 4.2) but not for $\delta^{13}C$ isotopes (ANOVA: $F_{1,99}$ = 1.37, P = 0.26; Fig. 4.2). Tukey multiple comparison tests showed that the differences between mean $\delta^{15}N$ isotope values resulted from higher $\delta^{15}N$ values in the tissues of eagles from Germany (13.6 ± 2.4 ‰) than in the tissues of eagles from Greenland (12.0 ± 1.9 ‰, P = 0.01). The isotope values of each population were distributed between the baseline data of isotopic signatures of marine and freshwater food sources (Fig. 4.2): German eagles exhibited lowest ^{13}C and highest ^{15}N enrichment, Finnish eagles showed intermediate ^{13}C and ^{15}N enrichment, and eagles from Greenland exhibited highest ^{13}C and lowest ^{15}N enrichment.

Seasonal and individual comparison

For the period 1996 to 2008, we analysed muscle samples of 75 white-tailed eagles from four federal states of Germany (Table 4.2). The range of $\delta^{13}C$ and $\delta^{15}N$ values in the German eagle population was very broad: -17.8 to -29.8 ‰ for $\delta^{13}C$ and 7.5 to 17.5 ‰ for $\delta^{15}N$ values. Since we did not detect significant differences between federal states (Kruskal-Wallis test, $\delta^{13}C$: χ^2 = 2.27, df = 3, P = 0.52; $\delta^{15}N$: χ^2 = 3.09, df = 3, P = 0.38) and years ($\delta^{13}C$: χ^2 = 4.50, df = 8, P = 0.81; $\delta^{15}N$: χ^2 = 13.3, df = 8, P = 0.10), we combined these data for further analysis.

Season had a significant effect on $\delta^{15}N$ ($F_{1,74}$ = 5.55, P = 0.02), with lower $\delta^{15}N$ values in the winter half-year than the summer half-year, but not on $\delta^{13}C$ ($F_{1,74}$ = 0.09, P = 0.77). The age of individuals significantly influenced $\delta^{13}C$ enrichment ($F_{1,74}$ = 4.54, P = 0.04), with lower $\delta^{13}C$ in juveniles than adults, but not $\delta^{15}N$ enrichment ($F_{1,74}$ = 1.24, P = 0.27). There was no significant differences between sexes for $\delta^{13}C$ values ($F_{1,74}$ = 0.89, P = 0.35) or $\delta^{15}N$ values ($F_{1,74}$ = 0.28, P = 0.60).

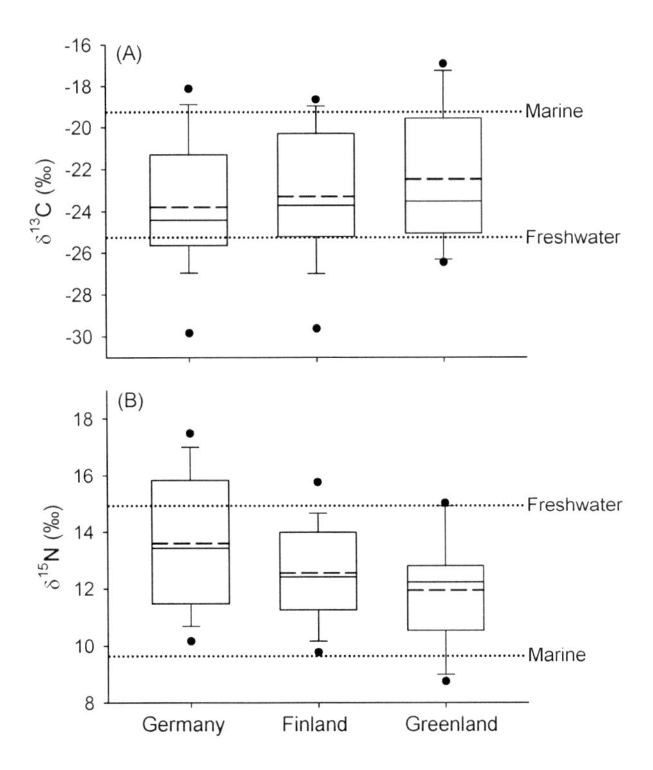

Fig. 4.2. *Stable (A) carbon and (B) nitrogen isotope ratios of white-tailed eagle populations inhabiting different countries: Germany, Finland, and Greenland. Data are presented as box plots with the ends of the box representing the 25th and 75th percentiles, horizontal lines showing 10th and 90th percentiles, and outermost points 5th and 95th percentiles. The solid and dashed lines within the boxes represent median and mean values, respectively. The dotted lines indicate mean isotopic signatures of marine and freshwater food sources (Mizutani et al. 1992; this study).*

Intra-individual comparison

A comparison between the isotopic signatures of liver and muscle tissues revealed significant differences in the German (Repeated measures MANOVA: Wilks' $\lambda = 0.78$, $F_{1,29} = 3.95$, $P = 0.03$) and Finnish (Wilks' $\lambda = 0.38$, $F_{1,9} = 6.31$, $P = 0.03$) white-tailed eagle population for combined isotopes and separately for $\delta^{13}C$ isotopes in German eagles (Repeated measures ANOVA: $F_{1,29} = 7.76$, $P = 0.009$) and for $\delta^{15}N$ isotopes in Finnish eagles ($F_{1,9} = 13.95$, $P = 0.006$). In contrast, the $\delta^{13}C$ and $\delta^{15}N$ values of liver and muscle tissues from Greenlandic eagles were very similar (Repeated measures MANOVA: Wilks' $\lambda = 1.0$, $F_{1,9} = 0.01$, $P = 0.99$).

Food sources

The isotope ratios for the 16 potential food species of the German white-tailed eagle population created a broad isotopic niche: mean $\delta^{13}C$ values ranged from -30.1 ‰ to -22.9 ‰ and mean $\delta^{15}N$ ranged from -0.5 ‰ to 16.2 ‰ (Table 4.1, Fig. 4.3). The $\delta^{15}N$ signatures discriminated significantly between species of aquatic and terrestrial environments (Mann-Whitney U-test: $U = 187.0$, $P < 0.0001$). Exclusively aquatic species exhibited the highest $\delta^{15}N$, predominantly aquatic species were intermediate and exclusively terrestrial species had the lowest $\delta^{15}N$ values. Piscivorous/carnivorous species were significantly enriched in ^{15}N in comparison to herbivorous/omnivorous species ($U = 440.0$, $P < 0.0001$). The $\delta^{13}C$ signatures also significantly discriminated between trophic levels, with lower $\delta^{13}C$ values in herbivorous species than in piscivorous/carnivorous species ($U = 62.0$, $P < 0.0001$).

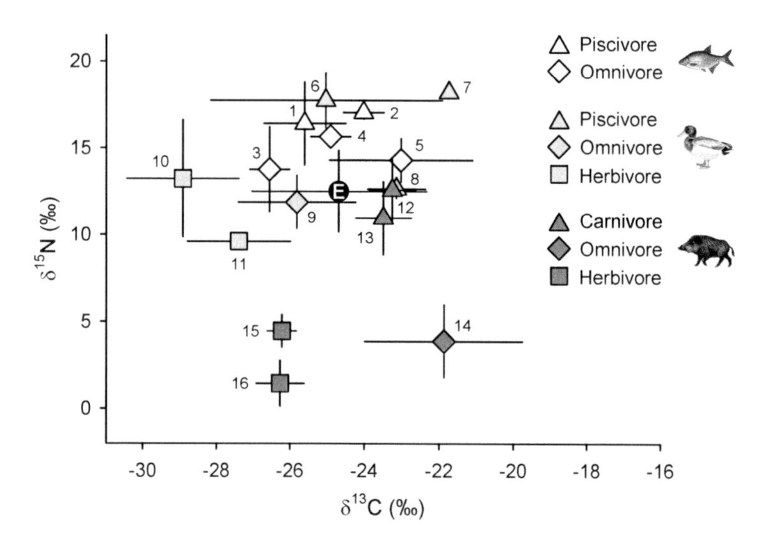

Fig. 4.3. Distribution of mean $\delta^{13}C$ and $\delta^{15}N$ muscle values of the German white-tailed eagle population and its potential food species (corrected for trophic discrimination) sampled in the federal state Mecklenburg-Western Pomerania in northeastern Germany. The eagle sample is represented by the filled circle with the letter "E"; fish, waterfowl and game mammal samples are depicted by white, light-grey and dark-grey symbols, respectively. Food species number labels are as detailed in Table 4.1. Error bars denote ± SD.

Table 4.2. *Summary of means, standard deviations (SD) and ranges (min to max) for $\delta^{13}C$ and $\delta^{15}N$ muscle values for the German white-tailed eagle population sampled in different federal states, seasons, age classes and sexes.*

	n	$\delta^{13}C$ [‰] (± SD)	Range [‰]	$\delta^{15}N$ [‰] (± SD)	Range [‰]
German white-tailed eagle population	75	-24.7 (± 2.4)	-17.8 to -29.8	12.5 (± 2.4)	7.5 to 17.5
Location					
Schleswig-Holstein	6	-25.2 (± 1.0)	-23.3 to -25.9	11.8 (± 2.3)	9.8 to 15.9
Mecklenburg-Western Pomerania	49	-24.9 (± 2.2)	-17.8 to -29.8	12.6 (± 2.4)	7.5 to 17.5
Brandenburg	14	-23.7 (± 3.3)	-18.9 to -29.8	13.0 (± 2.3)	9.8 to 17.4
Saxony	6	-25.2 (± 1.3)	-22.9 to -26.6	11.1 (± 2.3)	7.8 to 14.4
Season					
Winter half-year	55	-24.7 (± 2.5)	-17.8 to -29.8	12.1 (± 2.3)	7.5 to 17.5
Summer half-year	20	-24.8 (± 1.9)	-20.8 to -29.3	13.5 (± 2.3)	9.6 to 17.4
Age class					
Juvenile	36	-25.3 (± 2.1)	-17.8 to -29.8	12.2 (± 2.6)	7.5 to 17.4
Adult	39	-24.2 (± 2.5)	-18.4 to -29.6	12.8 (± 2.1)	7.8 to 16.5
Sex					
Male	35	-24.4 (± 2.7)	-17.8 to -29.8	12.7 (± 2.1)	9.8 to 16.5
Female	40	-24.3 (± 3.9)	-18.9 to -29.8	12.4 (± 2.6)	7.5 to 17.5

Table 4.3. *Mean $\delta^{13}C$ and $\delta^{15}N$ muscle values for and relative contribution of food sources to the white-tailed eagle diet in northeastern Germany. Percentages, standard errors (SE) and confidence limits of food source contributions were calculated using IsoError linear mixing models.*

Food source	n	$\delta^{13}C$ [‰] (± SD)	$\delta^{15}N$ [‰] (± SD)	Relative contribution in eagle diet [%] (SE; 95% confidence limits)	
				Summer half-year	Winter half-year
Aquatic	60	-	14.3 (± 2.9)	90.7 (7.7; 75.1 - 100)	73.4 (5.9; 61.8 - 85.1)
Terrestrial	30	-	6.2 (± 4.8)	9.3 (7.8; 0 - 24.9)	26.6 (5.9; 14.9 - 38.2)
Fish	29	-24.5 (± 1.4)	15.2 (± 1.5)	59.8 (24.0; 11.0 - 100)	47.8 (20.4; 7.4 - 88.2)
Waterfowl	31	-26.0 (± 1.2)	13.3 (± 4.2)	27.2 (28.5; 0 - 85.7)	22.7 (24.5; 0 - 71.3)
Game mammals	30	-23.8 (± 2.4)	6.2 (± 4.8)	13.0 (8.4; 0 - 29.9)	29.5 (6.9; 15.9 - 43.1)

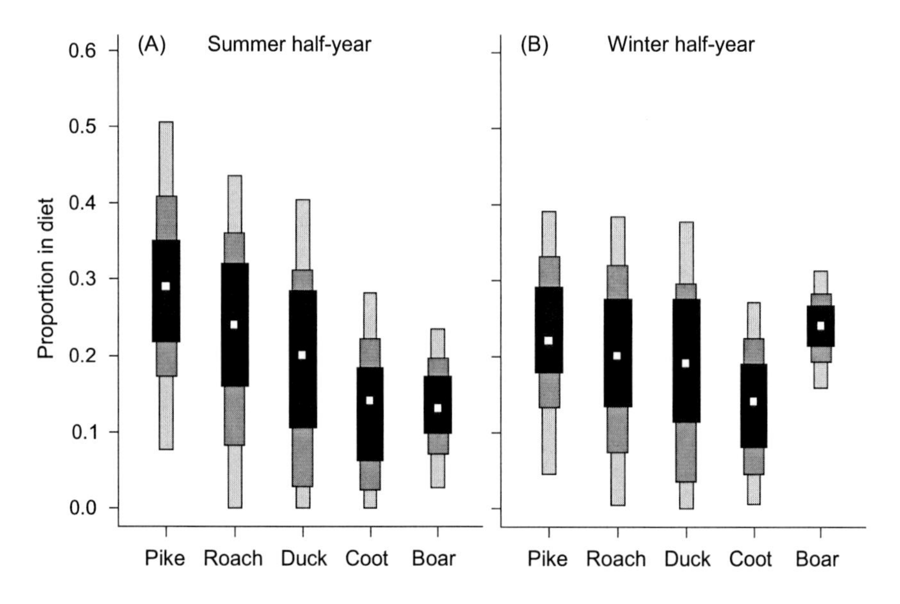

Fig. 4.4. *Proportions of five food sources in the diet of the German white-tailed eagle population predicted from stable isotope Bayesian mixing models (SIAR). Plot (A) and (B) show proportions for each food source in the summer half-year and the winter half-year, respectively. White squares represent the mean and dark-, medium-, and light-grey boxes indicate 50 %, 75 % and 95 % Bayesian credibility intervals.*

Eagle dietary mixing models

The IsoError two source linear mixing model, using $\delta^{15}N$ signatures, revealed that aquatic organisms accounted for 90.7 % and terrestrial organisms accounted for 9.3 % of the diet of the German white-tailed eagle population in the summer half-year (Table 4.3). The contribution of terrestrial food sources substantially increased, by 17.3 %, in the winter half-year. For each estimate, the SE and CI were relatively small (Table 4.3).

The IsoError three source linear mixing model, using both $\delta^{15}N$ and $\delta^{13}C$ signatures, showed that fish accounted for 59.8 %, waterfowl for 27.2 %, and game mammals for 13.0 % of eagle diet in the summer half-year (Table 4.3). The contribution of fish decreased, by 12 %, and the contribution of game mammals substantially increased, by 16.5 %, in the winter half-year. For each estimate, the SE and CI were relatively large (Table 4.3).

The Bayesian isotope mixing model SIAR generated the following 95 % credibility intervals of the posterior probability distributions of source proportions in the diet of the German white-tailed eagle population: pike accounted for between 8 % and 53 %, roach for between 0 % and 44 %, mallard duck for between 0 % and 41 %, coot for between 0 % and 28 %, and wild boar for between 3 % and 24 % in the eagle diet in the summer half-year (Fig. 4.4A). In the winter half-year, the proportions in the eagle diet decreased distinctively for pike (5-39 %) and roach (0-38 %), less so for duck (0-38 %) and coot (1-27 %). In contrast, the contributions of wild boar to the eagle diet substantially increased (16-31 %; Fig. 4.4B).

Discussion

The key results of the present study are (1) significant differences in the diet composition of white-tailed eagles between and within populations, (2) dietary variation and specialisation within individuals, and (3) seasonal dietary shifts resulting in important contributions of game mammals to the eagle diet. Since this study provides clear, quantitative evidence that game mammal carcasses can constitute an essential food source in white-tailed eagles, it has important implications for the conservation management of white-tailed eagles, particularly with respect to the adoption of non-toxic ammunition in game mammal hunting.

Dietary variation

As predicted, we found a large variation in the stable isotope ratios of white-tailed eagles. The spatial comparison of stable isotope ratios among populations from Germany, Finland and Greenland revealed significant differences between the German and Greenlandic populations.

This reflects the different foraging areas of the respective populations. German eagles forage predominantly on freshwater lakes and rivers (Fischer 1982), exhibited the highest ^{15}N and lowest ^{13}C enrichment and were thus closest to freshwater food sources. Finnish eagles obtain their food mainly from the brackish Baltic fringe (Stjernberg *et al.* 2003) and showed consequently an intermediate ^{15}N and ^{13}C enrichment. Greenlandic eagles forage mainly along the marine coast (Wille & Kampp 1983) and thus had the lowest ^{15}N and highest ^{13}C enrichment, indicating primarily marine food sources.

Our results provide evidence for dietary variation among populations stimulated by large-scale habitat and resource heterogeneity on the landscape level. Do individual white-tailed eagles from the same population exhibit the same feeding habits? If so, we would expect little isotopic variation among individuals, since feeding experiments with several bird species raised on an isotopically homogenous diet demonstrated so little variation in isotope ratios that it was limited to changes within the analytical error (± 0.2 ‰; Hobson & Clark 1992). The isotopic variance in the German white-tailed eagle population was much larger (SD > 2 ‰) and most likely represents dietary differences among free-ranging individual eagles. This finding agrees with the idea that inter-individual variation in feeding habits is responsible for the niche width of a predator population (Bolnick *et al.* 2003).

How does such variation arise among individuals of a population? We found significant differences of isotopic signatures among individuals sampled in the summer and winter half-year, which suggests that the diet composition of the German white-tailed eagle population varies between seasons. This finding is consistent with previous work documenting a functional response of white-tailed eagles to seasonal changes in food availability (M. Nadjafzadeh *et al.* unpublished data).

Furthermore, juvenile individuals exhibited a significantly lower ^{13}C and slightly lower ^{15}N enrichment than adults. It is unlikely that these lower isotopic ratios of juveniles emerge from age-related discrimination of heavy isotopes, as juvenile eagles are already full-grown, i.e. have the same body mass as adults (Glutz von Blotzheim *et al.* 1971). The foraging patterns of juvenile and adult eagles might diverge because adults not only have much more hunting experiences but they are also faster and more agile fliers than juveniles because of narrower and shorter remiges and rectrices (Glutz von Blotzheim *et al.* 1971). Moreover, available foraging space differs substantially between adults and juveniles: adult eagles are generally territorial, using home ranges with an average size of 15 km^2; observed extraterritorial movements of radio-tracked adults did not exceed a distance of 39 km from the centre of their home range (Scholz 2010). In contrast, juvenile eagles usually cover larger distances, particularly when they disperse from their natal area. Telemetry studies revealed mean travelling distances of 124 km per day and maximum travelling distances of 179 km per day (Nygård *et al.* 2003; M. Westphal *et al.* unpublished data). Accordingly, adult eagles have a better knowledge of the best foraging locations and may be better at capturing live prey than juveniles (M. Nadjafzadeh *et al.* unpublished data). Adults also appear more competent to hunt piscivorous prey species such as great crested grebes or cormorants which are highly vigilant and persistent prey. Juveniles may search more intensively for carcasses of large mammals such as deer during their flights and may focus on the capture of herbivorous prey-species such as coots which are easy to hunt and are known for poor diving capability (Struwe-Juhl 2003; M. Nadjafzadeh *et al.* unpublished data). Our findings are therefore consistent with observations suggesting higher carrion use by juvenile than adult white-tailed eagles (Halley & Gjershaug 2005) and studies documenting age-dependent differences in hunting skills and diet composition in raptorial birds (Edwards 1989; Rutz *et al.* 2006).

In spite of several previous studies on the diet of white-tailed eagles, information on age-related dietary differences was unavailable owing to the cryptic dispersal behaviour of juvenile eagles. Our results show that isotopic profiles of dietary variation among white-tailed eagles can reveal spatial, seasonal and individual differences and thus offer new insights into the feeding habits of free-ranging raptors.

Are white-tailed eagles dietary generalists or specialists?

We found high isotopic variance in white-tailed eagles both on the population and individual level. This poses the question whether these findings reflect individual generalisation or specialisation in a generalist predator. The latter seems unlikely because the isotopic analysis of liver and muscle tissues, deposited at different times and encompassing a time window of several weeks, revealed significant intra-individual differences in the isotopic composition of both the German and the Finnish white-tailed eagle population. Isotopic differences among tissues with different isotopic retention times reflect prey switching. In consequence, individuals of both populations represent isotopic and dietary generalists and occupy a relatively large niche width (Martínez del Rio *et al.* 2009).

In case of the Finnish eagles, the $\delta^{15}N$ values significantly varied, which may reflect the broad scope of different habitats, ranging from terrestrial habitats to freshwater, brackish water and marine habitats, and the associated wide variation in trophic levels occupied by the potential food species (Sulkava *et al.* 1997).

Individuals of the German eagle population varied significantly in $\delta^{13}C$ values, indicating higher variation in prey species than foraging habitats. This is plausible since German eagles pursue foraging tactics which include long-term rather than short-term shifts between foraging habitats, as they forage in the summer half-year mainly on freshwater lakes and switch to terrestrial foods sources during the winter half-year (M. Nadjafzadeh *et al.* unpublished data; this study).

However, the white-tailed eagle does not always occupy such a broad niche. The isotopic values of individuals of the Greenlandic eagle population, for example, were extremely similar among tissues, indicating a narrow niche width and a narrow, unchanging diet composition. This finding agrees with a study suggesting that Greenlandic eagles specialise on fish and is consistent with the fact that alternative food types to fish, such as waterfowl or game mammals, are scarce in Greenland owing to the unpredictable and harsh climate and icy conditions (Wille & Kampp 1983).

In conclusion, our findings revealed that white-tailed eagles vary in their feeding niches as a function of the breadth of habitats available, their associated food supplies and the seasonal changes therein, and reflecting individual generalisation as evidenced by dietary shifts or specialisation as a result of narrow, unchanging foraging patterns.

Eagle dietary mixing models

The IsoError two source mixing model revealed that aquatic organisms constitute the primary food source for the German white-tailed eagle population. Corresponding to the significant seasonal differences in ^{15}N enrichment, the contribution of terrestrial food sources substantially increased in the winter half-year, suggesting that terrestrial food sources are an important alternative diet in winter. Despite the variations in isotopic signatures both in eagles and food sources, we consider these results to be robust estimates, as the model produced small SEs and CIs, primarily due to the large differences between the nitrogen isotope ratios of food sources (Phillips & Gregg 2001).

The IsoError three source mixing model showed that fish constituted the major food item in the diet of German eagles. Waterfowl and game mammals were identified as alternative food types. In accordance with the finding of the two source mixing model, the contribution of fish decreased and the contribution of game mammal carcasses increased seasonally. The model produced high SEs and CIs because of the relatively small differences between source $\delta^{13}C$ signatures and the variations both within eagle and food source $\delta^{13}C$, $\delta^{15}N$ signatures (Phillips & Gregg 2001). Thus, the source proportion estimates should not be treated as absolute values but rather as relative trends which highlight seasonal dietary changes.

The estimates for the relative contribution of food species to the overall diet from the Bayesian mixing model SIAR revealed that white-tailed eagles primarily fed on pikes and roaches and supplemented their diet with mallard ducks, coots and wild boars. Again, there were clear changes over time with a decrease in the use of pikes and roaches throughout the winter half-year and a concurrent increase in the use of wild boar carcasses.

To conclude, the dietary mixing models indicate that white-tailed eagles of the German population act as generalist predators by seasonal switching to alternative terrestrial food types such as game mammal carrion. This dietary shift is probably the consequence of the decrease in availability and therefore the profitability of eagle primary aquatic prey (M. Nadjafzadeh *et al.* unpublished data). Our findings are in accordance with the alternative prey hypothesis which predicts that the assemblage of generalist predators such as foxes, mustelids, shrews, raptors and owls may respond to a decline in their main food types by shifting their diet to alternative food types (Reif *et al.* 2001).

The results for each dietary level are consistent with our classical diet study (M. Nadjafzadeh *et al.* unpublished data). We are therefore confident that stable isotope mixing models exhibit high explanatory power. They also offer substantial benefits over conventional methods for analysing raptor diets based on food remains and oral pellets because they provide an integration of the diet over periods of weeks and months, rather than being confined to represent the latest meal.

Implications for conservation

Our findings present robust quantitative evidences that white-tailed eagles use game mammal carcasses as important alternative food sources. The distinctive increase in the proportion of game mammals in the German eagle diet in the winter half-year corresponds to the main hunting season and a significant increase in the number of white-tailed eagles found lead-poisoned in Germany (Krone *et al.* 2009). These findings indicate that game mammals killed with lead-based ammunition constitute the major source of intoxication. The implication for the conservation management of at least the German white-tailed population is the removal of lead-based ammunition as an environmental pollutant that causes unnecessary and unwanted mortality. Flexible foraging tactics indicating carrion use also occurred in the Finnish population. Lead poisoning not only affects the German and Finnish populations but numerous other white-tailed eagle populations too (Kim *et al.* 1999; Krone *et al.* 2006; Helander *et al.* 2009). Lead poisoning is likely to be a serious threat also to many other scavenging raptors, including several species with high conservation priority (Fisher *et al.* 2006). Therefore, replacing lead-based ammunition by non-toxic alternatives in game mammal hunting would probably aid in the conservation of a wide range of raptor populations and species.

Acknowledgements

We thank the administration of the nature park *Nossentiner/Schwinzer Heide* and the *Reepsholt-Stiftung* for logistic support and accommodation. We are indebted to all people involved in the collection of white-tailed eagle carcasses, and our special thank goes to C. Herrmann, T. Stjernberg and F. Wille for organising the logistics of the carcass collection in Germany, Finland and Greenland, respectively. We are grateful to J. Fiebig for providing waterfowl muscle samples and to P. Wolf for providing game mammal muscle samples. We thank all cooperating fishermen and hunters for collaboration and support, and K. Blank and A. Trinogga for assistance and support. K. Sörgel performed the stable isotope ratio analysis. The study was funded by the Federal Ministry of Education and Research (BMBF, reference no. 0330720) and the Leibniz Institute for Zoo and Wildlife Research Berlin.

References

Angerbjörn, A., Hersteinsson, P., Lidén, K. & Nelson, E. 1994. Dietary variation in arctic foxes (*Alopex lagopus*) – an analysis of stable carbon isotopes. *Oecologia* 99: 226-232.

Austin, G.E., Thomas, C.J., Houston, D.C. & Thomson, D.B.A. 1996. Predicting the spatial distribution of buzzard *Buteo buteo* nesting areas using a Geographical Information System and remote sensing. *Journal of Applied Ecology* 33: 1541-1550.

Bolnick, D.I., Svanbäck, R., Fordyce, J.A., Yang, L.H., Davis, J.M., Hulsey, C.D. & Forister, M.L. 2003. The ecology of individuals: incidence and implications of individual specialization. *American Naturalist* 161: 1-28.

Caut, S., Roemer, G.W., Donlan, C.J. & Courchamp, F. 2006. Coupling stable isotopes with bioenergetics to estimate interspecific interactions. *Ecological Applications* 16: 1893-1900.

Caut, S., Angulo, E. & Courchamp, F. 2009. Variation in discrimination factors ($\Delta^{15}N$ and $\Delta^{13}C$): the effect of diet isotopic values and applications for diet reconstruction. *Journal of Applied Ecology* 46: 443-453.

Chamberlain, C.P., Waldbauer, J.R., Fox-Dobbs, K., Newsome, S.D., Koch, P.L., Smith, D.R., Church, M.E., Chamberlain, S.D., Sorenson, K.J. & Risebrough, R. 2005. Pleistocene to recent dietary shifts in California condors. *Proceedings of the National Academy of Sciences USA* 102: 16707-16711.

Cherel, Y., Hobson, K.A. & Weimerskirch, H. 2000. Using stable-isotope analysis of feathers to distinguish moulting and breeding origins of seabirds. *Oecologia* 122: 155-162.

DeNiro, M.J. & Epstein, S. 1978. Influence of diet on the distribution of carbon isotopes in animals. *Geochimica et Cosmochimica Acta* 42: 495-506.

DeNiro, M.J. & Epstein, S. 1981. Influence of diet on the distribution of nitrogen isotopes in animals. *Geochimica et Cosmochimica Acta* 45: 341-351.

Edwards, T.C. 1989. The ontogeny of diet selection in fledgling ospreys. *Ecology* 70: 881-896.

Fischer, W. 1982. *Die Seeadler*. Ziemsen Verlag, Wittenberg Lutherstadt, Germany.

Fisher, I.J., Pain, D.J. & Thomas, V.G. 2006. A review of lead poisoning from ammunition sources in terrestrial birds. *Biological Conservation* 131: 421-432.

Glutz von Blotzheim, U.N., Bauer, K.M. & Bezzel, E. 1971. *Handbuch der Vögel Mitteleuropas*. Volume 4: Falconiformes. Akademische Verlagsgesellschaft, Frankfurt am Main, Germany.

Halley, D.J. & Gjershaug, J.O. 1998. Inter- and intra-specific dominance relationships and feeding behaviour of golden eagles *Aquila chrysaetos* and sea eagles *Haliaeetus albicilla* at carcasses. *Ibis* 140: 295-301.

Hauff, P. 2008. Seeadler erobert weiteres Terrain. *Nationalatlas aktuell 1 (01/2008)*. Leibniz-Institut für Länderkunde [WWW document], Leipzig, Germany. URL http://nadaktuell.ifl-leipzig.de/Seeadler.1_01-2008.0.html

Helander, B. 1983. Reproduction of the white-tailed sea eagle *Haliaeetus albicilla* (L.) in Sweden, in relation to food and residue levels of organochlorine and mercury compounds in the eggs. *Ph.D. dissertation*, University of Stockholm, Stockholm, Sweden.

Helander, B., Axelsson, J., Borg, H., Holm, K. & Bignert, A. 2009. Ingestion of lead from ammunition and lead concentrations in white-tailed sea eagles (*Haliaeetus albicilla*) in Sweden. *Science of the Total Environment* 407: 5555-5563.

Hobson, K.A. 1999. Tracing origins and migration of wildlife using stable isotopes: a review. *Oecologia* 120: 314-326.

Hobson, K.A. & Clark, R.G. 1992. Assessing avian diets using stable isotopes II: factors influencing diet-tissue fractionation. *Condor* 94: 189-197.

Hunt, W.G., Burnham, W., Parish, C.N., Burnham, B., Mutch, B. & Oaks, J.L. 2006. Bullet fragments in deer remains: implications for lead exposure in scavengers. *Wildlife Society Bulletin* 34: 167-170.

Inger, R. & Bearhop, S. 2008. Applications of stable isotope analyses to avian ecology. *Ibis* 150: 447-461.

Kim, E.Y., Goto, R., Iwata, H., Masuda, Y., Tanabe, S. & Fujita, S. 1999. Preliminary survey of lead poisoning of Steller's sea eagle (*Haliaeetus pelagicus*) and white-tailed sea eagle (*Haliaeetus albicilla*) in Hokkaido, Japan. *Environmental Toxicology and Chemistry* 18: 448-451.

Krone, O., Stjernberg, T., Kenntner, N., Tataruch, F., Koivusaari, J., & Nuuja, I. 2006. Mortality, helminth burden and contaminant residues in white-tailed sea eagles from Finland. *Ambio* 35: 98-104.

Krone, O., Kenntner, N. & Tataruch, F. 2009. Gefährdungsursachen des Seeadlers (*Haliaeetus albicilla* L. 1758). *Denisia* 27: 139-146.

Martínez del Rio, C., Sabat, P., Anderson-Sprecher, R. & Gonzalez, S.P. 2009. Dietary and isotopic specialization: the isotopic niche of three *Cinclodes* ovenbirds. *Oecologia* 161: 149-159.

McCutchan, J.H. Jr., Lewis, W.M. Jr., Kendall, C. & McGrath, C.C. 2003. Variation in trophic shift for stable isotope ratios of carbon, nitrogen, and sulphur. *Oikos* 102: 378-390.

Mersmann, T.J., Buehler, D.A., Fraser, J.D. & Seegar, J.K.D. 1992. Assessing bias in studies of bald eagle food habits. *Journal of Wildlife Management* 56: 73-78.

Mizutani, H., Fukuda, M. & Kabaya, Y. 1992. ^{13}C and ^{15}N enrichment factors of feathers of 11 species of adult birds. *Ecology* 73: 1391-1395.

Moore, J.W. & Semmens, B.X. 2008. Incorporating uncertainty and prior information into stable isotope mixing models. *Ecology Letters* 11: 470-480.

Newsome, S.D., Collins, P.W., Rick, T.C., Guthrie, D.A., Erlandson, J.M. & Fogel, M.L. 2010. Pleistocene to historic shifts in bald eagle diets on the Channel Islands, California. *Proceedings of the National Academy of Sciences USA* 107: 9246-9251.

Nygård, T., Kenward, R.E. & Einvik, K. 2003. Dispersal in juvenile white-tailed sea eagles in Norway shown by radio-telemetry. In: *Sea eagle 2000*, (eds.) Helander, B., Marquiss, M. & Bowerman, B., pp. 191-196. Swedish Society for Nature Conservation/SNF, Stockholm, Sweden.

Oehme, G. 1975. Ernährungsökologie des Seeadlers, *Haliaeetus albicilla* (L.), unter besonderer Berücksichtigung der Population in den drei Nordbezirken der Deutschen Demokratischen Republik. *Doctoral dissertation*, Universität Greifswald, Greifswald, Germany.

Pain, D.J. & Amiard-Triquet, C. 1993. Lead poisoning of raptors in France and elsewhere. *Ecotoxicology and Environmental Safety* 25: 183-192.

Parnell, A.C., Inger, R., Bearhop, S. & Jackson, A.L. 2010. Source partitioning using stable isotopes: coping with too much variation. *PLoS ONE* 5: e9672.

Phillips, D.L. & Gregg, J.W. 2001. Uncertainty in source partitioning using stable isotopes. *Oecologia* 127: 171-179.

Phillips, D.L. & Gregg, J.W. 2003. Source partitioning using stable isotopes: coping with too many sources. *Oecologia* 136: 261-269.

Phillips, D.L., Newsome, S.D. & Gregg, J.W. 2005. Combining sources in stable isotope mixing models: alternative methods. *Oecologia* 144: 520-527.

Reif, V., Tornberg, R., Jungell, S. & Korpimäki, E. 2001. Diet variation of common buzzards in Finland supports the alternative prey hypothesis. *Ecography* 24: 267-274

Rutz, C., Whittingham, M.J. & Newton, I. 2006. Age-dependent diet choice in an avian top predator. *Proceedings of the Royal Society B* 273: 579-586.

Scholz, F. 2010. Spatial use and habitat selection of white-tailed eagles (*Haliaeetus albicilla*) in Germany. *Doctoral dissertation*, Freie Universität Berlin, Berlin, Germany.

Sinclair, A.R.E., Freyxell, J.M. & Caughley, G. 2006. *Wildlife ecology, conservation, and management*. 2nd edition. Blackwell Science, Oxford, UK.

Stjernberg, T., Koivusaari, J. & Högmander, J. 2003. Population trends and breeding success of the white-tailed sea eagle in Finland, 1970-2000. In: *Sea eagle 2000*, (eds.) Helander, B., Marquiss, M. & Bowerman, B., pp. 103-112. Swedish Society for Nature Conservation/SNF, Stockholm, Sweden.

Struwe-Juhl, B. 2003. Why do white-tailed eagles prefer coots? In: *Sea eagle 2000*, (eds.) Helander, B., Marquiss, M. & Bowerman, B., pp. 317-326. Swedish Society for Nature Conservation/SNF, Stockholm, Sweden.

Sulkava, S., Tornberg, R. & Koivusaari, J. 1997. Diet of the white-tailed eagle *Haliaeetus albicilla* in Finland. *Ornis Fennica* 74: 65-78.

Tieszen, L.L., Boutton, T.W., Tesdahl, K.G. & Slade, N.A. 1983. Fractionation and turnover of stable carbon isotopes in animal tissues: implications for $\delta^{13}C$ analysis of diet. *Oecologia* 57: 32-37.

Tieszen, L.L. & Boutton, T.W. 1988. Stable carbon isotopes in terrestrial ecosystem research. In: *Stable isotopes in ecological research*, (eds.) Rundel, P.W., Ehleringer, J.R., Nagy, K.A., pp. 167-195. Springer Verlag, Berlin, Germany.

Votier, S.C., Bearhop, S., MacCormick, A., Ratcliffe, N. & Furness, R.W. 2003. Assessing the diet of great skuas, *Catharacta skua*, using five different techniques. *Polar Biology* 26: 20-26.

Wille, F. 2003. Status of the white-tailed sea eagle in Greenland, 2000. In: *Sea eagle 2000*, (eds.) Helander, B., Marquiss, M. & Bowerman, B., pp. 27-29. Swedish Society for Nature Conservation/SNF, Stockholm, Sweden.

Wille, F. & Kampp, K. 1983. Food of the white-tailed eagle *Haliaeetus albicilla* in Greenland. *Holarctic Ecology* 6: 81-88.

CHAPTER 5

Who ingests what and how much? An experimental approach to simulate lead exposure and food processing of white-tailed eagles and other scavengers at shot mammalian carcasses

Abstract

Lead poisoning caused by feeding on shot mammalian carcasses containing fragments of lead-based bullets constitutes a serious hazard for avian scavengers. Not only obligate but also facultative scavenging species are affected, such as the white-tailed eagle *Haliaeetus albicilla*, for which lead intoxication is a major source of mortality. However, the exploitation of hunter-killed carrion by facultative avian scavenger communities is little studied. The mechanisms by what scavengers become intoxicated are unknown. As bullet fragments recovered from the intestinal tract of white-tailed eagles were generally small and did not represent the full spectrum of available bullet fragment sizes, the hypothesis arose that this is a result of selective food processing and eagles avoid the intake of large metal particles. Here, we used an experimental approach to detect lead-exposed species and explore the food processing of white-tailed eagles and other scavengers at shot mammalian carcasses. We conducted feeding experiments with free-ranging scavengers in six eagle home ranges at hunting glades in northeastern Germany, and with six eagles temporarily held in captivity. We provided ungulate carcasses containing non-toxic iron particles of different diameters, simulating bullet fragments, as food source. Primarily avian scavengers such as ravens, eagles and buzzards exploited the carcasses, indicating them to be exposed to lead fragments and poisoning. With increasing diameter of experimental particles, free-ranging scavengers and captive eagles increased the percentage of particles successfully avoided during feeding. Successful avoidance of particles was nearly perfect when the diameter of particles was 8.8 mm. Behavioural observations of captive eagles showed that they detected iron particles predominantly by touching the experimental carcass with their bill tip, suggesting that they use mechanoreceptors to judge food quality. Our findings indicate that conservation management of avian scavengers would be substantially improved if hunters used bullets that deform or fragment into particles greater than 9 mm in size to reduce the risk of metal ingestion and poisoning in white-tailed eagles and other birds exhibiting similar feeding behaviour. Since in contrast to conventional lead-based bullets, which fragment predominantly into small particles, lead-free bullets only deform or fragment into particles much larger than typical lead particles, they could constitute sustainable ammunition.

Keywords: avian scavengers, bullet fragments, *Haliaeetus albicilla*, lead poisoning, mechanoreception, probing behaviour, raptor conservation, selective food intake

Introduction

Lead poisoning has been documented as a major source of mortality in numerous avian scavenging species worldwide (Pain *et al.* 1995; Miller *et al.* 2002; Fisher *et al.* 2006). The main source of lead is suggested to be ungulates shot by hunters who sometimes leave carcasses or gut piles contaminated with fragments of lead-based bullets in the environment (Pain & Amiard-Triquet 1993; Kramer & Redig 1997; Hunt *et al.* 2006). Both obligate and facultative avian scavengers are affected by lead intoxication, such as the white-tailed eagle *Haliaeetus albicilla*, a facultative potential scavenger of mammalian carrion for which lead poisoning is a major cause of death (Krone *et al.* 2006; Krone *et al.* 2009*a*; Helander *et al.* 2009). The European populations of this raptor species are still in the process of recovery after severe population declines caused by human persecution, habitat degradation, and pollutants (Sulawa 2009).

Several studies documented that mammalian carrion represents an essential alternative food resource for facultative scavengers (DeVault 2003; Selva *et al.* 2003; Selva *et al.* 2005; Selva & Fortuna 2007; Cortés-Avizanda *et al.* 2009). However, they rarely considered the community of avian scavengers consuming hunter-killed ungulates or gut piles (Wilmers *et al.* 2003*b*; Blázquez *et al.* 2009) nor did they explore to what extent such carcasses are indeed the source of lead poisoning or by what mechanism scavengers are intoxicated. If avian scavengers are thought to accidentally ingest lead-based bullet fragments contained in shot mammalian carrion or gut piles, then the distribution of the sizes of lead-based bullet fragments recovered from poisoned birds should match the fragment sizes that can be recovered from shot carrion. Analyses of metal particles isolated from the intestinal tract of lead-poisoned eagles revealed that lead fragments from rifle bullets were the main cause of lead poisoning (Kenntner *et al.* 2001; Krone *et al.* 2006; Krone *et al.* 2009*a*), but recovered bullet fragments were generally very small and did not represent the full spectrum of bullet fragment sizes that develop after a bullet hits a shot mammal (Krone *et al.* 2009*b*). At present, it is unclear whether such a mismatch is (1) because eagles and other scavengers ingest lead-based bullet fragments from sources other than mammalian carrion, or (2) a consequence of selective feeding behaviour when processing contaminated mammalian carrion, in that eagles selectively remove or avoid the ingestion of larger metal fragments.

In a companion paper (M. Nadjafzadeh *et al.* unpublished data) we examine in detail the first hypothesis and its concomitant predictions, i.e. the importance of mammalian carrion as a food source for white-tailed eagles and the extent to which seasonal changes in availability and consumption of mammalian carrion match the temporal pattern observed in the occurrence of intoxicated birds. In this study, we used an experimental approach to examine (1) the use of shot mammalian carrion by different avian scavengers to identify those species potentially exposed to lead poisoning from mammalian carrion by having access to lead-based bullet fragments, (2) whether white-tailed eagles selectively avoid the intake of larger metal particles when feeding on mammalian carcasses, and (3) the sensory modality with which selective feeding behaviour may be facilitated. Raptors are known for their exceptional vision and therefore regarded to rely primarily on visual cues during feeding (Martin 1986). Except for neotropical vultures, their olfactory system is poorly developed (Bang & Cobb 1968; Houston 1986). In contrast to other vertebrates, avian taste organs apparently exist only in small numbers (Mason & Clark 2000). Like many birds, raptors have tactile somatosensory receptors in their bill (Gottschaldt 1985; Necker 2000) but most studies examining their function during food intake focused on shorebirds (Dias *et al.* 2009; Cunningham *et al.* 2009) and therefore, nothing is currently known, whether and how raptors may use mechanoreception during food processing. If white-tailed eagles detect metal particles in their food by visual and/or mechanical cues, then their capability to avoid such particles should increase with particle size. Alternatively, if eagles rely predominantly on chemical stimuli to detect metal particles, they should not distinguish between particles of different sizes.

Material and methods

Study site, captive facilities and study animals

Data were collected from 2007 to 2009. Feeding experiments with free-ranging scavengers were carried out during the main hunting seasons, from September to January, in the nature park *Nossentiner/Schwinzer Heide* in Mecklenburg-Western Pomerania, northeastern Germany (53°30'-53°40'N, 11°59'-12°35'E; Fig. 5.1). This area is inhabited by one of Germany's highest population densities of white-tailed eagles with 15 breeding pairs on 356 km^2. The landscape is characterised by numerous water bodies, high amounts of secondary forest and agriculture and low amounts of settlement and traffic. We performed experiments in the home ranges of six individually identified territorial eagle pairs (Scholz 2010).

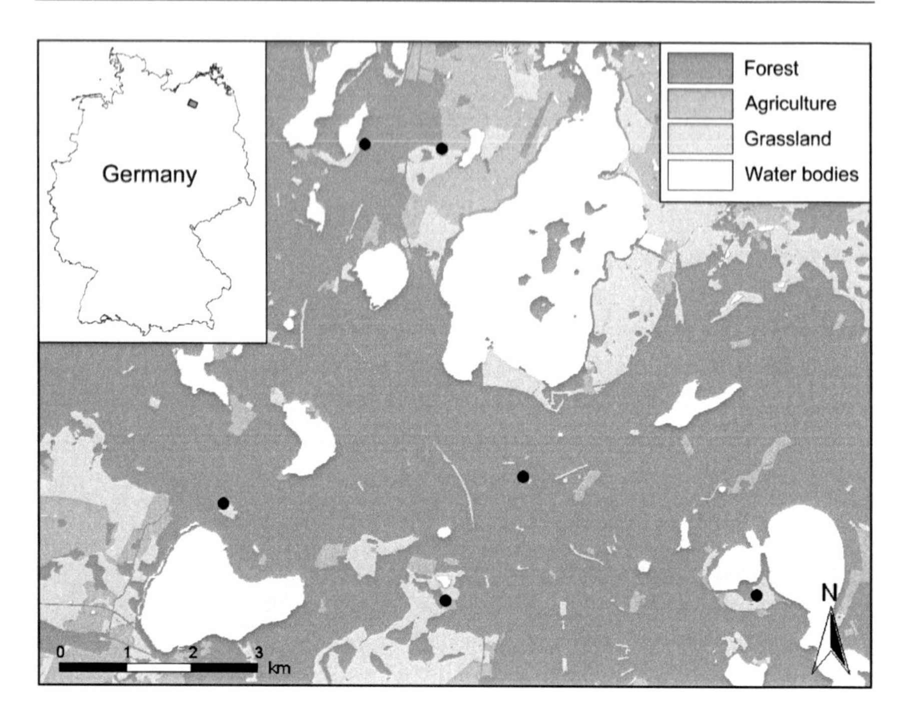

Fig. 5.1. *Map of the study area in the nature park Nossentiner/Schwinzer Heide located in northeastern Germany. The locations where feeding experiments were carried out are shown as filled circles (N = 6).*

Feeding experiments with captive white-tailed eagles were conducted at the nature conservation station *Woblitz*, located in Brandenburg, northeastern Germany. *Woblitz* temporarily accommodates injured free-ranging raptors for rehabilitation. Our study animals were six eagles (Table 5.1) housed separately in octagonal flight aviaries (6 m high x 16 m diameter). Except for a few spyholes, the aviary walls were cladded with timber or steel plates in order to reduce anthropogenic disturbance. The aviary top was made of a wide-meshed net for illumination by ambient daylight. All study animals were successfully released back into the wild after the conclusion of our feeding experiments.

Table 5.1. *Identities of six white-tailed eagle individuals and the period held in captivity before feeding experiments started.*

Study animal	Sex	Age class	Period in captivity
M1	male	adult[a]	307 days
M2	male	immature[b]	155 days
F1	female	adult	12 days
F2	female	adult	28 days
F3	female	adult	41 days
F4	female	subadult[c]	17 days

[a]> 5 years old; [b]2 years old; [c]4 years old

Experimental design

Experiments with free-ranging scavengers were designed to study the composition of the avian scavenger community and gain insight into their natural feeding behaviour on shot game left in the field. Since field conditions did not allow us to identify which scavengers ingested or avoided iron particles, we carried out feeding experiments with eagles temporarily held in captivity to study in detail their ability to selectively avoid or remove experimental particles. We provided scavengers carcasses of wild ungulates or gut piles containing iron particles as food source. Smooth iron particles instead of bullet fragments or lead shot were used to prevent risk of injury or toxicological exposure to the study animals (Hapke 1988; Fiedler & Rösler 1993; Kelly *et al.* 1998; Mitchell *et al.* 2001; Brewer *et al.* 2003).

The experimental carcasses were prepared with smooth iron particles of standardised sizes by distributing them regularly over each carcass. Carcasses were either prepared with (1) iron balls to simulate small lead fragments and lead pellets, or (2) iron nuts to simulate larger rifle bullet parts. To examine whether the particle shape has an influence on feeding behaviour, we used balls and nuts of similar sizes. Table 5.2 summarizes the diameters and numbers of balls and nuts used during each experimental feeding trial. We conducted one trial per eagle home range with balls and one trial with nuts, respectively, in the field experiments, and three trials with balls and three trials with nuts, respectively, per captive white-tailed eagle. During field experiments, food intake by free-ranging scavengers was observed by a remote video surveillance system, to minimise disturbance of feeding. During feeding trials with captive white-tailed eagles, food intake was documented by a hidden observer using focal animal sampling combined with continuous recording (Martin & Bateson 1993).

Table 5.2. *Diameters and numbers of iron balls and nuts presented during each feeding experiment to free-ranging scavengers and captive white-tailed eagles.*

Diameter [mm]		Free-ranging scavengers		Captive white-tailed eagles	
		Number		Number	
Balls	Nuts	Balls	Nuts	Balls	Nuts
1.0	2.7	3	2	1	1
2.0	4.4	3	2	1	1
3.0	6.0	3	2	1	1
4.0	7.7	3	2	1	1
5.0	8.8	3	2	1	1

In cooperation with local hunters, we chose open hunting glades as experimental sites for the field experiments, where hunter-killed mammalian carrion or gut piles would be regularly found under natural circumstances (Fig. 5.1). A mobile and weather-proof video surveillance system (VSS) for carcass monitoring was used, containing a day & night colour camera (MTV-63 X11 P-K, Lechner electric CCTV, Kolbermoor, Germany), digital video recorder (DVR, DiREX-Pro.A30, BWA Technology GmbH, Münster, Germany), movement detector (ES 90, Roos Electronics GmbH, Arnsberg, Germany), and infrared illumination (IR spotlight 940nm, Securitec-Gerlach, Schwarme, Germany). We camouflaged the VSS and mounted the camera and the movement detector on brackets on trees at the experimental sites. At dawn, the experimental carcasses were placed approximately 10 meters away from the VSS and secured to the ground to prevent the carcass being moved outside the area monitored by the surveillance system.

We defined an iron particle as avoided if the supplied carcass was completely consumed by study animals and only bones and isolated iron particles remained at the site. The experimental sites in both the field experiments and the flight aviaries were examined after each trial with the help of a metal detector (CS 660, C-Scope, Ashford, UK) to localise any remaining iron particles on the ground.

To investigate the sensory modalities responsible for possible selective feeding, we observed in detail the probing behaviour prior to intake of food items of each captive eagle. We distinguished (1) 'viewing', defined as detection of iron particles by visual cues, followed by direct, apparently purposeful (without repeated touching of the carcass tissue) mechanical separation of the particle from the experimental carcass tissue with the bill or the bill tip; (2) 'touching', defined as detection of iron particles by tactile cues perceived with the bill tip if eagles repeatedly probed the experimental carcass tissue with the bill tip followed by disengaging the bill from the touched location and then touching, grabbing and swallowing other parts of the experimental carcass; (3) 'disgorging', defined as detection of iron particles by tactile cues perceived within the bill in that eagles grabbed the experimental carcass tissue with the beak without prior repeated touching followed by spitting out a particle.

Data analysis

We analysed the video records from the field experiments using the software Interact (Mangold International GmbH, Arnstorf, Germany). From these records we calculated the relative frequency of occurrence of each avian and mammalian scavenger species by dividing the number of trials where a given species was present by the total number of trials.

We also calculated the species-specific 'absolute carrion intake' of each scavenger species of each trial c_{kl} (in grams) as $c_{kl} = d_{kl} \times f_k$, where d_{kl} is the 'scavenging duration' (in minutes) of species k during trial l and f_k the 'active feeding rate' of species k (in g min^{-1}). The 'scavenging duration' d_{kl} was defined as the sum of the measured feeding times (in minutes) over all individuals for species k during trial l. 'Active feeding rates' f_k were determined in the following way. First, we provided the captive white-tailed eagles with pre-weighed gut piles or muscle tissue and counted the number of pecks per feeding bout to determine the average 'bite size' (in grams) as the amount consumed by an eagle per peck (Table 5.3). Average 'bite size' per peck of ravens (*Corvus corax*) and magpies (*Pica pica*) were derived from Wilmers *et al.* (2003*a*). Second, we measured peck rates (per minute) for white-tailed eagles, ravens and magpies by randomly choosing focal animals from the video records at the experimental feeding sites and counting peck rates as number of pecks min^{-1} at the carcasses. Third, 'bite size' per peck was multiplied by peck rates to determine f_k in grams per minute (Table 5.3). The 'active feeding rate' f_k is therefore a short-term rate of food intake averaged over one feeding bout, in contrast to intake rates labelled as 'consumption rates' which are long-term rates of food intake averaged over a longer time period such as one day (Wilmers *et al.* 2003*a*). Since f_k is a positive function of species body mass (Travaini *et al.* 1998), we calculated f_k for other scavenger species present at our feeding trials by using linear interpolation (Table 5.3). Finally, the species-specific 'relative carrion intake' of species k in trial l was calculated by dividing the absolute carrion intake c_{kl} by the sum of absolute carrion intakes across all species in trial l.

Table 5.3. *Determination of feeding parameters to estimate feeding rates f_k for scavenger species k present at experimental feeding sites. Data are means ± standard errors of mean (SE).*

Scavenger species	Body mass [g][a]	Bite size [g] per peck (SE)	Pecking rate [min^{-1}] (SE)	f_k [g min^{-1}]
White-tailed eagle	5200	3.41 (0.34)	14.08 (0.15)	47.98
Red kite[b]	1078	-	-	22.82
Common buzzard[b]	985	-	-	20.65
Common raven	1250	1.15 (0.40)[d]	23.33 (0.18)	26.83
Hooded crow[b]	560	-	-	10.73
Black-billed magpie	206	0.09 (0.004)[d]	27.68 (0.20)	2.49
Eurasian jay[b]	177	-	-	1.82
Raccoon dog[c]	6962	-	-	82.48
Red fox[c]	6125	-	-	66.09

[a]Average body mass values for raptor, corvid, and canid species are from Glutz von Blotzheim *et al.* (1971), Glutz von Blotzheim & Bauer (1993), and Niethammer & Krapp (1993), respectively.

[b]Estimated from linear interpolation based on the values of f_k for white-tailed eagles, ravens and black-billed magpie and average species-specific body mass.

[c]Estimated from linear interpolation based on the values of f_k for white-tailed eagles and coyotes (*Canis latrans*; Wilmers & Stahler 2002) and average species-specific body mass.

[d]Values are from Wilmers *et al.* (2003*a*).

The relative frequency of avoidance of each captive eagle j for particles of the ith size category was calculated as $F_{ij} = P_{ij} / n_{ij}$, where P_{ij} is the number of avoided particles of size category i in eagle j and n_{ij} the total number of particles available of size category i in trials with eagle j ($n_{ij} = \sum_{i=1}^{n} P_{ij}$). In the field experiments, we did not differentiate between individuals or species but different feeding trials and calculated the relative frequency of avoidance F_{il}, where P_{il} represents the number of avoided particles of size category i in trial l, and n_{il} the total number of particles available of the size category i in trial l. For the purpose of comparing the relative frequency of particle avoidance between the two types of experiments (captive eagles and free-ranging scavengers), the number of avoided and ingested iron particles was pooled over the feeding trials of the free-ranging scavengers and the captive eagles, respectively.

To investigate how particle size influenced the ability to feed selectively, we compared the proportion of iron particles of different size categories available in the experimental carcasses with the proportion of iron particles of different size categories ingested during feeding by calculating the selection ratio $\hat{w}_i$ (Manly *et al.* 2002). In the case of the captive eagle feeding trials, the selection ratio $\hat{w}_i$ for the *i*th particle size category is $\hat{w}_i = u_{i+} / \sum_{j=1}^{n} \pi_{ij} u_{+j}$, where u_{i+} is the number of particles of size category i ingested by all captive eagles, π_{ij} is the proportion of particles of size category i available to captive eagle j, and u_{+j} is the number of particles of size category i ingested by captive eagle j. In the case of the field experiments, we did not distinguish between individuals or species but between feeding trials and calculated $\hat{w}_i$ such that u_{i+} represented the number of particles of size category i ingested by scavengers in all trials, π_{il} the proportion of particles of size category i available to scavengers in trial l, and u_{+l} the number of particles of size category i ingested by scavengers in trial l. The selection ratio $\hat{w}_i$ can range from 0 to ∞, with values < 1 indicating ingestion less frequently than expected and values > 1 indicating ingestion more frequently than expected from the particles' respective availabilities.

Statistical analysis

To assess differences between arrival times of avian and mammalian scavengers at field experimental feeding sites, we used Mann-Whitney U-tests. We employed the log-likelihood ratio test to compare frequencies of iron particle avoidance between and within free-ranging scavengers and captive white-tailed eagles. Changes in the avoidance of particles with size were tested with the Friedman test. To test whether the frequency of particle avoidance in free-ranging scavengers and captive eagles differed between balls and nuts, we applied the Wilcoxon signed-ranks test. We also used this test to compare the frequency of occurrence of probing behaviour categories 'touching' and 'disgorging' in the six captive eagles. The selection ratios $\hat{w}_i$ of free-ranging scavengers and captive eagles for each particle size category i were tested for significant deviation from 1 by comparing $\{(\hat{w}_i - 1) / \mathrm{SE}(\hat{w}_i)\}^2$ with critical values for the χ^2 distribution with one degree of freedom (Manly *et al.* 2002). We used linear regression models to estimate the relationship between the frequency of particle avoidance ($F_{ij/l}$) and particle size in free-ranging scavengers and captive eagles. All statistical tests were two-tailed and carried out with SPSS 16.0 (SPSS Inc., Chicago, IL, USA). The significance level was set at $P < 0.05$; for multiple comparisons, we used the Bonferroni procedure to adjust significance levels, i.e. we divided the nominal significance level 0.05 by the number of pairwise comparisons (Manly *et al.* 2002). Unless otherwise indicated, data are presented as means ± standard deviations (SD).

Ethical note

Although we designed our experiments in such a way that no risk of injury or toxicological exposure seemed possible for white-tailed eagles and all other animals scavenging on the experimental carcasses, we kept the number of trials and the dosages of particles to the minimum necessary to conduct statistical comparisons. In order to increase sample size for the purpose to identify lead-exposed species, we conducted in four white-tailed eagle home ranges one additional feeding experiment, respectively, with unprepared experimental carcasses. Captive eagles observed in this study were not caught and held in captivity for our experiments, but were present for other reasons such as convalescence from injuries. The experiments were permitted by the relevant government animal welfare authorities in Brandenburg and Mecklenburg-Western Pomerania and approved by the animal welfare officer at the Leibniz Institute for Zoo and Wildlife Research Berlin.

Results

Patterns of carcass consumption

We provided 16 mammalian carcasses, 12 prepared with iron particles and four unprepared, to scavengers in six white-tailed eagle home ranges of the study area (Fig. 5.1). Since the carrion use by present scavenger species was similar at prepared and unprepared experimental carcasses (see Appendix, Table 5.A1), we refer in the following to summarised results of all conducted feeding trials (Table 5.4). The carcasses were completely consumed after 25.8 ± 20.4 h. Seven avian and two mammalian species exploited the experimental carcasses (Table 5.4). Common ravens, white-tailed eagles, and common buzzards (*Buteo buteo*) were the major scavengers, present at 69 to 100 % of the provided carcasses with a mean 'scavenging duration' of between 55 and 128 min and a mean 'relative carrion intake' of between 19 and 35 % (Table 5.4). Hooded crows (*Corvus corone cornix*) were also frequent scavengers, present at 75 % of the carcasses, with a mean 'scavenging duration' of 34 min and a mean

Table 5.4. *Use of experimental carcasses by scavengers observed via remote video surveillance in 16 feeding trials in six white-tailed eagle home ranges in the nature park Nossentiner/Schwinzer Heide.*

Scavenger species	% trials with species present	'scavenging duration' [min]	'absolute carrion intake' [g]	'relative carrion intake' [%]
White-tailed eagle	93.8	55.1 ± 87.1	2643.9 ± 4176.7	31.5 ± 30.4
Adult[a]	68.8	23.2 ± 38.5	1111.6 ± 1846.4	15.2 ± 24.5
Juvenile[b]	75.0	31.9 ± 78.1	1532.3 ± 3747.4	16.3 ± 26.7
Common buzzard	68.8	74.1 ± 132.5	1499.1 ± 2734.0	19.0 ± 32.1
Red kite	12.5	3.1 ± 8.6	70.5 ± 195.5	0.5 ± 1.4
Common raven	100	127.5 ± 212.4	3383.1 ± 5721.6	35.3 ± 24.8
Hooded crow	75.0	33.8 ± 58.9	362.2 ± 632.2	7.9 ± 11.3
Black-billed magpie	25.0	11.9 ± 36.3	29.5 ± 90.3	0.8 ± 2.4
Eurasian jay	12.5	0.03 ± 0.1	0.05 ± 0.21	0.002 ± 0.008
Red fox	62.5	2.8 ± 6.9	187.8 ± 458.2	3.8 ± 8.4
Raccoon dog	12.5	0.7 ± 2.2	44.2 ± 176.7	1.1 ± 4.5

[a]> 5 years old; [b]≤ 1-5 years old

'relative carrion intake' of 8 %. Red foxes (*Vulpes vulpes*) were present in 63 % of the trials, but scavenged on average only 3 min and ingested approximately 4 % of the carcasses. Minor scavengers included red kites (*Milvus milvus*), black-billed magpies, Eurasian jays (*Garrulus glandarius*) and raccoon dogs (*Nyctereutes procyonoides*).

The first scavengers arrived after 2.1 ± 1.4 h at the feeding site and were always birds, predominantly common ravens (in 10 of 16 trials, 62.5 %). Common buzzards were the first arrivals at the carcass in three cases, hooded crows in two cases and white-tailed eagles in one case. Mammals arrived significantly later than birds (Mann-Whitney U-test, $U = 21.0$, $P = 0.002$, $N = 28$); red foxes visited the carcasses after 18.2 ± 20.6 h and raccoon dogs after 15.0 ± 2.5 h. In 50 % of the visits by red foxes and raccoon dogs, the experimental carcasses were already totally consumed. The white-tailed eagles visiting and feeding on the experimental carcasses included both adult territory owners and roaming juvenile eagles (Table 5.4).

Selective food processing

The free-ranging scavengers and the captive white-tailed eagles did not ingest all iron particles hidden in the experimental carcasses. The percentage of iron balls whose ingestion during food processing was avoided comprised 18.9 ± 8.8 % and 21.1 ± 5.0 %, and the percentage of iron nuts avoided was 43.3 ± 10.3 % and 50.0 ± 7.0 % for free-ranging scavengers and captive eagles, respectively. There were no significant differences in the relative frequency of avoidance between these two test groups (log-likelihood ratio test, iron balls: $G = 0.14$, $P = 0.71$, $N = 180$; iron nuts: $G = 0.64$, $P = 0.43$, $N = 150$), between the trials with free-ranging scavengers in six eagle home ranges ($G = 4.99$, $df = 1$, $P = 0.44$, $N = 150$) or between the six individuals of captive eagles ($G = 1.67$, $df = 1$, $P = 0.89$, $N = 180$).

Size influenced the relative frequency with which iron particles were avoided in both free-ranging scavengers and captive white-tailed eagles (Friedman test, free-ranging scavengers, balls: $\chi^2 = 21.2$, $df = 4$, $P = 0.0003$, $N = 6$; nuts: $\chi^2 = 21.1$, $df = 4$, $P = 0.0003$, $N = 6$; captive eagles, balls: $\chi^2 = 21.6$, $df = 4$, $P = 0.0002$, $N = 6$; nuts: $\chi^2 = 21.7$, $df = 4$, $P = 0.0002$, $N = 6$). Free-ranging scavengers and captive eagles exhibited the same selective feeding pattern: balls of 1.0 mm, 2.0 mm and 3.0 mm diameter were ingested at significantly higher proportions than expected from their respective availability, balls of 4.0 mm diameter were ingested according to their availability, and balls of 5.0 mm diameter were ingested at significant lower proportion than expected from their availability (Table 5.5). Nuts of 2.7 mm and 4.4 mm diameter were ingested more often than expected, nuts of 6.0 mm diameter were ingested as expected, and nuts of 7.7 mm and 8.8 mm diameter were ingested less often than expected from their availability. Free-ranging scavengers and captive white-tailed eagles displayed a similar degree of selective food processing, as expressed by the minimum and maximum observed proportion of ingested iron particles (Table 5.5).

In both test groups, the frequency of particle avoidance did not differ significantly between balls and nuts of diameters 3.0 mm and 2.7 mm, 4.0 mm and 4.4 mm, and 5.0 mm and 6.0 mm, respectively (Wilcoxon signed-ranks tests, all $P_{exact} > 0.09$, $N = 6$).

The relative frequency of particle avoidance linearly increased with particle size (free-ranging scavengers: adj. $r^2 = 0.80$, $F_{1,9} = 35.8$, $P = 0.0003$; captive eagles: adj. $r^2 = 0.85$, $F_{1,9} = 50.9$, $P < 0.0001$; Fig. 5.2). Iron nuts of 8.8 mm diameter were almost completely avoided, with 91.7 ± 20.4 % in free-ranging scavengers and 94.5 ± 13.6 % in captive eagles.

Table 5.5. *The influence of the size and shape of iron particles hidden in experimental carcasses on their chance of being ingested by (A) free-ranging scavengers in the nature park Nossentiner/Schwinzer Heide, and (B) captive white-tailed eagles.* o_i, *SE* o_i, $o_{il/j}$*min,* $o_{il/j}$*max: proportion (and its standard error, minimum and maximum values) of iron particle i ingested;* π_i*: proportion of available iron particle i;* $\hat{w}_i$, *SE* $\hat{w}_i$*: selection ratio (and its standard error); D: direction of behaviour, +: over-ingested, –: avoided, 0: ingested according to expectation based on availability.*

Diameter of iron particles	o_i	SE o_i	o_{il}min	o_{il}max	π_i	$\hat{w}_i$	SE $\hat{w}_i$	χ^2	P	D
A) Free-ranging scavengers										
Balls										
1.0 mm	0.25	0.05	0.21	0.27	0.20	1.23	0.06	14.63	0.00013	+
2.0 mm	0.25	0.05	0.21	0.27	0.20	1.23	0.06	14.63	0.00013	+
3.0 mm	0.25	0.05	0.21	0.27	0.20	1.23	0.06	14.63	0.00013	+
4.0 mm	0.19	0.05	0.17	0.21	0.20	0.96	0.09	0.92	0.34	0
5.0 mm	0.07	0.03	0.00	0.14	0.20	0.34	0.14	35.29	< 0.0001	–
Nuts										
2.7 mm	0.35	0.08	0.29	0.50	0.20	1.76	0.13	33.92	< 0.0001	+
4.4 mm	0.32	0.08	0.25	0.40	0.20	1.62	0.09	46.94	< 0.0001	+
6.0 mm	0.21	0.07	0.17	0.29	0.20	1.03	0.12	0.06	0.80	0
7.7 mm	0.09	0.05	0.00	0.17	0.20	0.44	0.19	8.82	0.003	–
8.8 mm	0.03	0.03	0.00	0.14	0.20	0.15	0.14	36.91	< 0.0001	–

Table 5.5. *continued*

Diameter of iron particles	o_i	SE o_i	o_{ij}min	o_{ij}max	π_i	$\hat{w}_i$	SE $\hat{w}_i$	χ^2	P	D
B) Captive white-tailed eagles										
Balls										
1.0 mm	0.25	0.05	0.23	0.27	0.20	1.27	0.03	66.08	< 0.0001	+
2.0 mm	0.25	0.05	0.23	0.30	0.20	1.27	0.03	66.08	< 0.0001	+
3.0 mm	0.24	0.05	0.18	0.27	0.20	1.20	0.06	10.95	0.0009	+
4.0 mm	0.18	0.05	0.17	0.23	0.20	0.92	0.05	2.39	0.12	0
5.0 mm	0.07	0.03	0.00	0.09	0.20	0.35	0.07	96.62	< 0.0001	–
Nuts										
2.7 mm	0.38	0.07	0.25	0.50	0.20	1.89	0.17	26.72	< 0.0001	+
4.4 mm	0.36	0.07	0.29	0.43	0.20	1.78	0.09	67.34	< 0.0001	+
6.0 mm	0.18	0.06	0.11	0.25	0.20	0.89	0.13	0.72	0.40	0
7.7 mm	0.07	0.04	0.00	0.14	0.20	0.33	0.14	22.65	< 0.0001	–
8.8 mm	0.02	0.02	0.00	0.11	0.20	0.11	0.10	69.32	< 0.0001	–

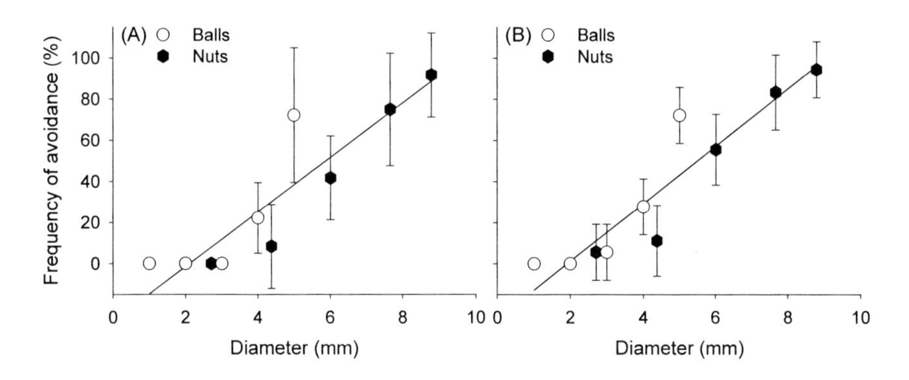

Fig. 5.2. Relationship between the avoidance of iron particles during feeding and iron particle size by (A) free-ranging scavengers in the nature park Nossentiner/Schwinzer Heide and (B) captive white-tailed eagles. The x-axis shows the diameter of particles, the y-axis shows the mean percentage of particles avoided in (A) 12 feeding trials with free-ranging scavengers and (B) 36 feeding trials with six captive eagles. Error bars denote ± SD.

Probing behaviour of experimentally modified food items

We never observed that the captive white-tailed eagles detected iron particles in their food only by visual cues. We observed 'touching', i.e. detecting of iron particles with the bill tip, in all six focal eagles and 'disgorging', i.e. detecting of iron particles within the beak, in four of six eagles (M2, F1, F2 and F3). The relative frequency of occurrence of touching for detecting and avoiding of iron particles during food processing was significantly higher than the relative frequency of disgorging, with focal eagles using touching in 92.3 ± 7.2 % and disgorging in 7.7 ± 7.2 % of occasions (Wilcoxon signed-ranks test, $P_{exact} = 0.03$, N = 6). The behaviour patterns of the eagles did not significantly change between the avoidance of balls or nuts ($P_{exact} = 0.88$, N = 6; Fig. 5.3).

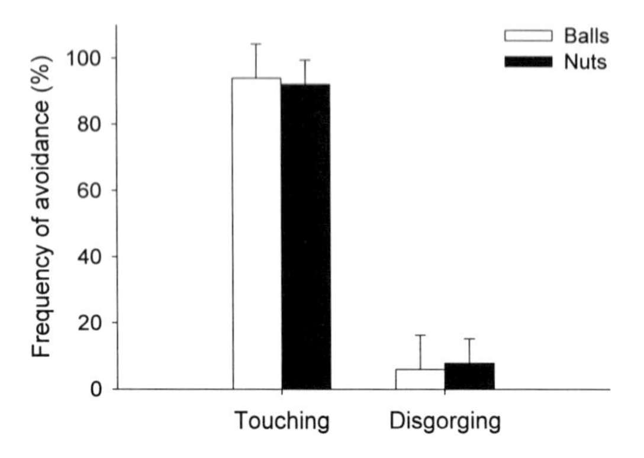

Fig. 5.3. *Frequency of occurrence of the two categories of probing behaviour 'touching' and 'disgorging' in six captive white-tailed eagles for detecting iron balls and iron nuts during food processing. Error bars denote ± SD.*

Discussion

Lead exposure in avian scavengers

At hunting glades, where carcasses and gut piles shot by hunters usually occur during the hunting season, avian species were the major scavengers at our experimental carcasses. This is in agreement with a study from North America where bald eagles (*Haliaeetus leucocephalus*) and ravens dominated the consumption at hunter kills (Wilmers *et al.* 2003*b*) and a finding in the Białowieża Primeval Forest (eastern Poland) that carcasses in open habitats were more easily detected by avian scavengers (Selva *et al.* 2003). Birds are highly suitable to detect carrion owing to their superiority of visual resolution over that of other animals (Martin 1986) and the lower cost of foraging effort in soaring locomotion in comparison to running (Schmidt-Nielson 1972). It is therefore likely that their search efforts cover large areas more efficiently than mammalian scavengers and may have contributed to the detection success recorded in this study. However, our experimental design suited avian scavengers because we put out the carcasses at dawn, similar to hunters who shoot their game at the same time. It does not, however, reflect likely access by nocturnal mammalian scavengers such as red foxes or raccoon dogs to carcasses shot by hunters early in the evening or during the night and which would first become accessible to them. So, the 'relative carrion intake' recorded in this study for each species will reflect the intake of scavengers active during daylight but may underrepresent the true intake by nocturnal scavengers across all carcasses shot by hunters.

Common ravens were the scavengers to arrive first and most frequently at the carcasses and to exhibit the longest scavenging duration with the highest carrion intake. This could be a consequence of social foraging by juvenile ravens which form communal roosts and that serve as information centres during autumn and winter (Wright *et al.* 2003). Thus, hunter-killed carrion seems to represent a frequent food resource for ravens and we expect them to be a species highly exposed to lead. This is in accordance with the very high lead concentrations in blood documented in common ravens with access to big-game offal (Craighead & Bedrosian 2008). Consistent with the frequent incidence of lead poisoning in white-tailed eagles (Kenntner *et al.* 2001; Krone *et al.* 2009*a*; Helander *et al.* 2009), the eagles were the second most important scavengers after ravens and thus play a major role in scavenger guilds. This suggests that hunter-killed mammalian carrion constitutes a noticeable component of the diet of white-tailed eagles – and this has been confirmed by a recent study on the feeding ecology of white-tailed eagles in the same study area (M. Nadjafzadeh *et al.* unpublished data). Our observations documented that not only the territorial adult eagle pairs but also roaming adolescent eagles exploited the experimental carcasses and were therefore exposed to lead poisoning. This is consistent with findings that mortality caused by lead poisoning affected white-tailed eagles of all age classes except nestlings (Sulawa 2009). The third important avian scavenger was the common buzzard. In these raptors, lead poisoning has also been reported in several countries (Pain & Amiart-Triquet 1993; Pain *et al.* 1995).

This study demonstrates that the frequency of presence of a species may only be loosely linked to its relative carrion intake. This could be a consequence of later arrival times at the carcasses as in the case of red foxes or lower 'active feeding rates' as in the case of the competitively inferior hooded crow. Short-term records of the presence of scavengers at a carcass are therefore inappropriate to provide a representative record of the absolute or relative species-specific carrion intake. The only appropriate method to accurately determine these parameters appear at present to be the continuous observation of feeding sites via remote video surveillance. In order to identify the probability of exposure to lead in scavenger communities, the ideal method would be a combination of remote video surveillance as applied here with the individual identification of scavenger individuals to estimate what proportion of the local population of a scavenging species visits lead-contaminated carcasses.

Avoidance of large metal particles by avian scavengers

Our feeding experiments provide the first evidence that scavengers avoid the ingestion of large metal particles during feeding. Although we do not know the extent to which each species selectively avoided the intake of large metal particles, it is likely that the scavengers primarily feeding on the experimental carcasses, the common ravens, white-tailed eagles and common buzzards, all showed this behaviour. This would also be consistent with the fact that the community of free-ranging scavengers and the captive eagles did not differ in any obvious way in their manner of food processing. The experiments with the captive eagles confirmed for this species active, selective avoidance of large particles. The similarity of results across all captive eagles suggests that our results were not influenced by conditions in captivity, or the duration of stay in captivity, and are representative for this species. A biological basis for such selective avoidance is the observation that the majority of birds, including eagles, ravens, buzzards and even vulture species, tend to avoid the ingestion of hard, undigestible parts of their food such as bone fragments (Rea 1973; Houston 1986; Houston 1988; Heinrich 1988; Rosenberg & Cooper 1990; Mersmann *et al.* 1992; Tornberg & Reif 2007; Slagsvold *et al.* 2010).

Captive eagles were slightly more selective and displayed a lower variance in the avoidance of metal particles than the free-ranging scavenger community. We do not think that this is a methodical artefact, although it is possible that with our method the detection of iron particles in the field left behind by free-ranging scavengers was more difficult than in flight aviaries. The differences in selectivity and its variance between free-ranging and captive conditions could be a consequence of varying species-specific abilities to selectively avoid the intake of large metal particles. Or it could be the consequence of (1) a nutritional state in free-ranging scavengers that is on average worse than that of captive ones, thereby reducing selectivity during feeding (Mueller 1973; Hansen 1986), and/or (2) poor foraging skills of juvenile scavengers, leading to increased risk-taking and decreased selectivity (East 1988). Additionally, inter- and intraspecific competition may lead to a faster food intake and thus influence the feeding behaviour of subordinate species or individuals (East 1988; Halley & Gjershaug 1998).

As expected, the avoidance of particles depended on their size. Iron particles up to 3 mm were completely ingested by the free-ranging scavengers and almost completely consumed by the captive white-tailed eagles. Avoidance was almost complete by both test groups for particles with a diameter of 8.8 mm and this did not depend on their shape. It is therefore particle size rather than particle shape which influences the degree to which hard, indigestible food items are selectively avoided during food processing.

Probing behaviour and sensory modalities

Since avian scavengers such as white-tailed eagles can avoid the ingestion of large particles, which sensory modalities are responsible for this ability to selectively process food? As we found a linear relationship between selectivity of avoidance and particle size, the predominant use of chemical senses is unlikely. The olfactory system of eagles is likely to be poor (Bang & Cobb 1968) and therefore unlikely to be sufficiently sensitive to detect olfactory stimuli from metal particles inserted deeply inside carrion above the intensive smell that emanates from carrion itself (Houston 1986). In addition, the probing behaviour 'disgorging', which would be compatible with gustation, was only seen in four of the six focal animals and rarely observed. This is consistent with the lack of gustatory papillae on the tongue of white-tailed eagles (Jackowiak & Godynicki 2005). Many waterfowl species are known to selectively ingest lead shot as grit (Mateo *et al.* 2000), making it unlikely that there is a general aversiveness in birds to metallic tasting particles.

Surprisingly, we never observed that the focal eagles detected the iron particles by vision only. Are therefore tactile receptors in the bill tip, or perhaps receptors inside the whole beak crucial to the detection of unwanted food items? Our results strongly support the first possibility. All focal eagles reacted primarily to tactile cues perceived with their bill tip to avoid the ingestion of metal particles during feeding. The bill tip of raptors is an important tool, equivalent to human fingertips, for handling food, including activities such as plucking hairs, feathers and scales of their prey, grabbing pieces of flesh from bones and skins, or feeding recently hatched chicks with tiny bites of food (Houston 1988; Necker 2000; Slagsvold *et al.* 2010). In several avian species that, like raptors, use their beak as an instrument to catch, select or manipulate food (e.g., parrots or waterfowl), a complex sensory structure in the bill tip, the "bill tip organ", has been described (Gottschaldt 1985). The observation that focal eagles disgorged particles suggest that independent of gustation, the mechanoreceptors such as Grandy and Herbst corpuscles distributed on the inside of the upper and lower jaw (Gottschaldt 1985) can be sensitive enough to detect particles eagles do not want to swallow if they were already in the beak.

Overall, our findings indicate that raptors have sensitive mechanoreceptors in their beak facilitating the avoidance of indigestible particles such as metal fragments during food intake.

This suggests that studies investigating sensory modalities associated with foraging and feeding should consider senses other than the apparently most important one, as vision would be in the case of raptors, in order to understand the mechanics of each step of food processing.

Implications for conservation

White-tailed eagles and several other avian scavengers such as common buzzards and common ravens were the first to arrive and had the highest 'relative carrion intake", i.e. they were the main scavengers, at hunting glades. These species are therefore highly susceptible to lead poisoning during the hunting season, when carcasses containing lead shot or lead bullet fragments became an abundant food source. Since iron particles up to 3 mm diameter were almost completely ingested, our results demonstrate that lead shot as well as lead fragments resulting form conventional rifle bullets will be ingested by scavengers. On the other hand, white-tailed eagles and probably also buzzards and ravens selectively avoided the ingestion of larger metal particles, and particles with a diameter of 8.8 mm were almost completely avoided. Thus, lead-free rifle bullets that only deform or fragment into pieces greater than 9 mm could substantially reduce the risk of metal ingestion and poisoning in avian scavengers. The implications for the conservation management of scavenging birds are therefore (1) the removal of lead-based ammunition to prevent unnecessary mortality in birds, and (2) the adoption of already existing lead-free rifle bullets (Rosenberger 2007). In contrast to the conventional lead-core rifle bullets that highly fragment into particles of varying but predominantly small sizes, lead-free rifle bullets either deform and keep 90-100 % of their original mass, or fragment into particles much larger than typical lead-based fragments (Rosenberger 2007). Accordingly, particles from lead-free bullets are suitable to be avoided by eagles during food intake. Such lead-free bullets would therefore meet the requirements based on our findings to provide hunters sustainable alternative ammunition for game mammal hunting.

Acknowledgements

We are grateful to the administration of the nature park *Nossentiner/Schwinzer Heide*, the nature conservation station *Woblitz* and the *Reepsholt-Stiftung* for their collaboration and support. We are indebted to K. Eichhorn for preparing the video surveillance system and to F. Scholz, A. Trinogga and N. Kenntner for assistance and support. We thank all cooperating forestry districts, hunters and farmers in the study area. The study was funded by the Federal Ministry of Education and Research (BMBF, reference no. 0330720) and the Leibniz Institute for Zoo and Wildlife Research Berlin, and administrated by Project Management Juelich (PtJ).

References

Bang, B.G. & Cobb, S. 1968. The size of the olfactory bulb in 108 species of birds. *Auk* 85: 55-61.

Blázquez, M., Sánchez-Zapata, J.A., Botella, F., Carrete, M. & Eguía S. 2009. Spatio-temporal segregation of facultative avian scavengers at ungulate carcasses. *Acta Oecologia* 35: 645-650.

Brewer, L., Fairbrother, A., Clark, J. & Amick, D. 2003. Acute toxicity of lead, steel, and an iron-tungsten-nickel shot to mallard ducks (*Anas platyrhynchos*). *Journal of Wildlife Diseases* 39: 638-648.

Cortés-Avizanda, A., Selva, N., Carrete, M. & Donázar, J.A. 2009. Effects of carrion pulses on herbivore spatial distribution are mediated by facultative scavengers. *Basic and Applied Ecology* 10: 265-272.

Craighead, D. & Bedrosian, B. 2008. Blood lead levels of common ravens with access to big-game offal. *Journal of Wildlife Management* 72: 240-245.

Cunningham, S.J., Castro, I. & Potter, M.A. 2009. The relative importance of olfaction and remote touch in prey detection by North Island brown kiwis. *Animal Behaviour* 78: 899-905.

DeVault, T.L., Rhodes, O.E., Shivik, Jr. & Shivik, J.A. 2003. Scavenging by vertebrates: behavioral, ecological, and evolutionary perspectives on an important energy transfer pathway in terrestrial ecosystems. *Oikos* 102: 224-234.

Dias, M.P., Granadeiro, J.P. & Palmeirim, J.M. 2009. Searching behaviour of foraging waders: does feeding success influence their walking? *Animal Behaviour* 77: 1203-1209.

East, M. 1988. Crop selection, feeding skills and risks taken by adult and juvenile rooks *Corvus frugilegus*. *Ibis* 130: 294-299.

Fiedler, H.J. & Rösler, H.J. 1993. *Spurenelemente in der Umwelt*. Gustav Fischer Verlag, Jena, Germany.

Fisher, I.J., Pain, D.J. & Thomas, V.G. 2006. A review of lead poisoning from ammunition sources in terrestrial birds. *Biological Conservation* 131: 421-432.

Glutz von Blotzheim, U.N., Bauer, K.M. & Bezzel, E. 1971. *Handbuch der Vögel Mitteleuropas*. Volume 4: Falconiformes. Akademische Verlagsgesellschaft, Frankfurt am Main, Germany.

Glutz von Blotzheim, U.N. & Bauer, K.M. 1993. *Handbuch der Vögel Mitteleuropas*. Volume 13: Passeriformes. Aula-Verlag, Wiesbaden, Germany.

Gottschaldt, K.-M. 1985. Structure and function of avian somatosensory receptors. In: *Form and function in birds*. Volume 3, (eds.) King, A.S. & McLelland, J., pp. 375-463. Academic Press, London, UK.

Halley, D.J. & Gjershaug, J.O. 1998. Inter- and intra-specific dominance relationships and feeding behaviour of golden eagles *Aquila chrysaetos* and sea eagles *Haliaeetus albicilla* at carcasses. *Ibis* 140: 295-301.

Hansen, A.J. 1986. Fighting behavior in bald eagles: a test of game theory. *Ecology* 67: 787-797.

Hapke, H.-J. 1988. *Toxikologie für Veterinärmediziner*. Ferdinand Enke Verlag, Stuttgart, Germany.

Heinrich, B. 1988. Winter foraging at carcasses by three sympatric corvids, with emphasis on recruitment by the raven, *Corvus corax*. *Behavioral Ecology and Sociobiology* 23: 141-156.

Helander, B., Axelsson, J., Borg, H., Holm, K. & Bignert, A. 2009. Ingestion of lead from ammunition and lead concentrations in white-tailed sea eagles (*Haliaeetus albicilla*) in Sweden. *Science of the Total Environment* 407: 5555-5563.

Houston, D.C. 1979. The adaptations of scavengers. In: *Serengeti, dynamics of an ecosystem,* (eds.) Sinclair, A.R.E. & Griffiths, M.N., pp. 263-286. University of Chicago Press, Chicago, USA.

Houston., D.C. 1986. Scavenging efficiency of turkey vultures in tropical forest. *Condor* 88: 318-323.

Houston, D.C. 1988. Competition for food between Neotropical vultures in forest. *Ibis* 130: 402-417.

Hunt, W.G., Burnham, W., Parish, C.N., Burnham, B., Mutch, B. & Oaks, J.L. 2006. Bullet fragments in deer remains: implications for lead exposure in scavengers. *Wildlife Society Bulletin* 34: 167-170.

Jackowiak, H. & Godynicki, S. 2005. Light and scanning electron microscopic study of the tongue in the white tailed eagle (*Haliaeetus albicilla*, Accipitridae, Aves). *Annals of Anatomy* 187: 251-259.

Kelly, M.E., Fitzgerald, S.D., Aulerich, R.J., Balander, R.J., Powell, D.C., Stickle, R.L., Stevens, W., Cray, C., Tempelman, R.J. & Bursian, S.J. 1998. Acute effects of lead, steel, tungsten-iron, and tungsten-polymer shot administered to game-farm mallards. *Journal of Wildlife Diseases* 34: 673-687.

Kenntner N., Tataruch, F. & Krone, O. 2001. Heavy metals in soft tissue of white-tailed eagles found dead or moribund in Germany and Austria from 1993 to 2000. *Environmental Toxicology and Chemistry* 20: 1831-1837.

Kramer, J.L. & Redig, P.T. 1997. Sixteen years of lead poisoning in eagles, 1980-95: an epizootiologic view. *Journal of Raptor Research* 31: 327-332.

Krone, O., Stjernberg, T., Kenntner, N., Tataruch, F., Koivusaari, J., & Nuuja, I. 2006. Mortality, helminth burden and contaminant residues in white-tailed sea eagles from Finland. *Ambio* 35: 98-104.

Krone, O., Kenntner, N. & Tataruch, F. 2009*a*. Gefährdungsursachen des Seeadlers (*Haliaeetus albicilla* L. 1758). *Denisia* 27: 139-146.

Krone, O., Kenntner, N., Trinogga, A., Nadjafzadeh, M., Scholz, F., Sulawa, J., Totschek, K., Schuck-Wersig, P. & Zieschank, R. 2009*b*. Lead poisoning in white-tailed sea eagles: causes and approaches to solutions in Germany. In: *Ingestion of lead from spent ammunition: implications for wildlife and humans*, (eds.) Watson, R.T., Fuller, M., Pokras, M. & Hunt, G., pp. 289-301. The Peregrine Fund, Idaho, USA.

Manly, B.F.J., McDonald, L.L., Thomas, D.L. & McDonald, T.L. 2002. *Resource selection by animals*. 2nd edition. Kluwer Academic Publishers, Dordrecht, Netherlands.

Martin, G.R. 1986. Shortcomings of an eagle's eye. *Nature* 319: 357.

Martin, P. & Bateson, P. 1993. *Measuring behaviour*. 2nd edition. Cambridge University Press, Cambridge, UK.

Mason, J.R. & Clark, L. 2000. The chemical senses in birds. In: *Strukie's avian physiology*. 5th edition, (ed.) Whittow, G.C., pp. 39-56. Academic Press, London, UK.

Mateo, R., Guitart, R. & Green, A.J. 2000. Determinants of lead shot, rice and grit ingestion in ducks and coots. *Journal of Wildlife Management* 64: 939-947.

Mersmann, T.J., Buehler, D.A., Fraser, J.D. & Seegar, J.K.D. 1992. Assessing bias in studies of bald eagle food habits. *Journal of Wildlife Management* 56: 73-78.

Miller, M.J.R., Wayland, M.E. & Bortolotti, G.R. 2002. Lead exposure and poisoning in diurnal raptors: a global perspective. In: *Raptors in the New Millenium,* (eds.) Yosef, R., Miller, M.L. & Pepler, D., pp. 224-245. International Birding & Research Center, Eilat, Israel.

Mitchell, R.R., Fitzgerald, S.D., Aulerich, R.J., Balander, R.J., Powell, D.C., Tempelman, R.J., Stickle, R.L., Stevens, W. & Bursian, S.J. 2001. Health effects following chronic dosing with tungsten-iron and tungsten-polymer shot in adult game-farm mallards. *Journal of Wildlife Diseases* 37: 451-458.

Mueller, H.C. 1973. The relationship of hunger to predatory behaviour in hawks (*Falco sparverius* and *Buteo platypterus*). *Animal Behaviour* 21: 513-520.

Necker, R. 2000. The somatosensory system. In: *Strukie's avian physiology*. 5th edition, (ed.) Whittow, G.C., pp. 57-70. Academic Press, London, UK.

Niethammer, J. & Krapp, F. 1993. *Handbuch der Säugetiere Europas*. Volume 5: Raubsäuger – Carnivora. Aula-Verlag, Wiesbaden, Germany.

Pain, D.J. & Amiard-Triquet, C. 1993. Lead poisoning of raptors in France and elsewhere. *Ecotoxicology and Environmental Safety* 25: 183-192.

Pain, D.J., Sears, J. & Newton, I. 1995. Lead concentrations in birds of prey in Britain. *Environmental Pollution* 87: 173-180.

Rea, A.M. 1973. Turkey vultures casting pellets. *Auk* 90: 209-210.

Rosenberg, K.V. & Cooper, R.J. Approaches to avian diet analysis. 1990. *Studies in Avian Biology* 13: 80-90.

Rosenberger, M.R. 2007. *Jagdgeschosse*. Motorbuchverlag, Stuttgart, Germany.

Schmidt-Nielson, K. 1972. Energy cost of swimming, running and flying. *Science* 177: 222-228.

Scholz, F. 2010. Spatial use and habitat selection of white-tailed eagles (*Haliaeetus albicilla*) in Germany. *Doctoral dissertation*, Freie Universität Berlin, Berlin, Germany.

Selva, N., Jędrzejewska, B. & Jędrzejewski, W. 2003. Scavenging on European bison carcasses in Białowieża Primeval Forest (eastern Poland). *Ecoscience* 10: 303-311.

Selva, N., Jedrzejewska, B. & Jedrzejewski, W. 2005. Factors affecting carcass use by a guild of scavengers in European temperate woodland. *Canadian Journal of Zoology* 83: 1590-1601.

Selva, N. & Fortuna, M.A. 2007. The nested structure of a scavenger community. *Proceedings of the Royal Society B* 274: 1101-1108.

Slagsvold, T., Sonerud, G.A., Grønlien, H.E. & Stige, L.C. 2010. Prey handling in raptors in relation to their morphology and feeding niches. *Journal of Avian Biology* 41: 488-497.

Sulawa, J. 2009. Impact of lead poisoning on the dynamics of a recovering population: a case study of the german white-tailed eagle (*Haliaeetus albicilla*) population. *Doctoral dissertation*, Freie Universität Berlin, Berlin, Germany.

Tornberg, R. & Reif, V. 2007. Assessing the diet of birds of prey: a comparison of prey items found in nests and images. *Ornis Fennica* 84: 21-31.

Travaini, A., Donázar, A.J., Rodríguez, A., Ceballos, O., Funes, M., Delibes, M. & Hiraldo, F. 1998. Use of European hare (*Lepus europaeus*) carcasses by an avian scavenging assemblage in Patagonia. *Journal of Zoology* 246: 175-181.

Wilmers, C.C. & Stahler, D.R. 2002. Constraints on active-consumption rates in gray wolves, coyotes and grizzly bears. *Canadian Journal of Zoology* 80: 1256-1261.

Wilmers, C.C., Crabtree, R.L., Smith, D.W., Murphy, K.M. & Wayne, M.G. 2003*a*. Trophic facilitation by introduced top predators: grey wolf subsidies to scavengers in Yellowstone National Park. *Journal of Animal Ecology* 72: 909-916.

Wilmers, C.C., Stahler, D.R., Crabtree, R.L., Smith, D.W. & Wayne, M.G. 2003*b*. Resource dispersion and consumer dominance: scavenging at wolf- and hunter-killed carcasses in greater Yellowstone, USA. *Ecology Letters* 6: 996-1003.

Wright, J., Stone, R.E. & Brown, N. 2003. Communal roosts as structured information centres in the raven, *Corvus corax*. *Journal of Animal Ecology* 72: 1003-1014.

Appendix

Table 5.A1. *Use of experimental carcasses by scavengers observed via remote video surveillance in (A) 12 feeding trials providing carcasses prepared with iron particles in six white-tailed eagle home ranges and (B) four feeding trials providing unprepared carcasses in four white-tailed eagle home ranges in the nature park Nossentiner/Schwinzer Heide.*

Scavenger species	% trials with species present	'scavenging duration' [min]	'absolute carrion intake' [g]	'relative carrion intake' [%]
A) Use of prepared experimental carcasses				
White-tailed eagle	100	60.7 ± 96.2	2913.5 ± 4615.3	31.9 ± 31.9
Adult[a]	75.0	20.9 ± 32.3	1002.1 ± 1551.0	14.6 ± 22.7
Juvenile[b]	75.0	39.8 ± 89.6	1911.4 ± 4297.3	17.3 ± 30.3
Red kite	8.33	1.73 ± 5.98	39.4 ± 136.4	0.19 ± 0.66
Common buzzard	66.7	70.2 ± 128.7	1408.0 ± 2654.2	18.4 ± 34.9
Common raven	100	141.8 ± 242.7	3754.0 ± 6541.5	34.5 ± 26.5
Hooded crow	66.7	40.0 ± 66.0	429.2 ± 707.8	9.37 ± 12.5
Black-billed magpie	25.0	15.7 ± 41.6	39.1 ± 103.5	1.10 ± 2.75
Eurasian jay	8.33	0.04 ± 0.13	0.07 ± 0.24	0.003 ± 0.009
Raccoon dog	16.7	0.89 ± 2.49	58.9 ± 204.1	1.51 ± 5.22
Red fox	66.7	1.36 ± 3.24	90.2 ± 214.1	3.05 ± 8.13
B) Use of unprepared experimental carcasses				
White-tailed eagle	75.0	38.2 ± 58.8	1835.2 ± 2821.1	30.3 ± 29.6
Adult[a]	50.0	30.0 ± 59.1	1440.1 ± 2834.2	17.1 ± 33.3
Juvenile[b]	75.0	8.23 ± 9.39	395.1 ± 450.5	13.2 ± 13.5
Red kite	25.0	7.18 ± 14.4	163.8 ± 327.7	1.33 ± 2.65
Common buzzard	75.0	85.8 ± 163.6	1772.5 ± 3377.8	21.0 ± 26.4
Common raven	100	84.6 ± 79.8	2270.4 ± 2140.6	37.8 ± 22.2
Hooded crow	100	15.0 ± 28.0	161.4 ± 299.9	3.59 ± 5.68
Black-billed magpie	25.0	0.31 ± 0.62	0.77 ± 1.53	0.006 ± 0.012
Eurasian jay	25.0	0	0	0
Raccoon dog	0	0	0	0
Red fox	50.0	6.77 ± 13.2	480.6 ± 854.1	5.99 ± 9.93

[a]> 5 years old; [b]≤ 1-5 years old

CHAPTER 6

General discussion

This dissertation was designed to gather detailed information about the foraging ecology and feeding habits of white-tailed eagles. The implementation of complementary methods identified the most effective approach for dietary investigations on free-ranging raptor species. The results provided new insights into the diet composition, foraging patterns and food preferences of white-tailed eagles. The findings revealed that the foraging strategies of white-tailed eagles are influenced by numerous factors and correspond to the main predictions of the optimal prey choice model. The present study offered an answer concerning the main sources of lead poisoning and presented an approach for the solution of the lead problem in white-tailed eagles and other scavenging birds.

Advantages and drawbacks of different approaches for raptor diet analysis

In terms of methodology, I intended to carry out a comparison of several approaches to questions regarding raptor feeding ecology. I tested whether they yield similar results and what type of information can be obtained in order to identify the most promising technique for future dietary research on raptors. The results showed that all methods applied had advantages and drawbacks:

(1) I used classical approaches such as the collection of prey remains and pellets and the analysis of stomach contents of white-tailed eagles which allowed the identification of diet components down to species level. The combination of these complementary methods minimised potential biases in diet estimates arising from each technique (Simmons *et al.* 1991; Mersmann *et al.* 1992; Redpath *et al.* 2001). In contrast to most raptor diet studies, this study used a systematic and comprehensive collection of prey remains and pellets. This was possible because of detailed information about feeding sites and home range borders of individually identified territorial white-tailed eagle pairs obtained from direct observations and telemetry tracking. Hence, (i) the typical source of error concerning the overestimation of food types close to the eyries or of individual diets was minimised. Repeated data collection prevented that (ii) the data represented only snapshots of feeding habits or were biased towards periods with a high probability of finding a large amount of food remains such as the breeding season (Tjernberg 1981; Sulkava *et al.* 1997). Furthermore, a regular sampling in monthly intervals reduced (iii) the overrepresentation of persistent remains such as bones and

the underrepresentation of ephemeral remains such as gut piles. However, the collection of food remains and pellets was extremely time- and labor-intensive with regard to the output of detailed data on "only" seven white-tailed eagle pairs.

The investigation of stomach contents from white-tailed eagles all over Germany provided a representative profile of the feeding habits of the German white-tailed eagle population, but was also time-consuming, as the collection of 126 eagle carcasses with intact and filled stomachs comprised a time span of eight years. To summarise, the main advantage of the combination of classical approaches applied in the present study was the gain of a very detailed dietary knowledge down to individual food species of individual eagles. These results can be regarded as reliable estimates, since the systematic sampling protocol minimised the potential sources of error. The main drawback is the required expenditure of time and (wo)man-power for data acquisition, limiting the collection of long-term data to a relatively small number of individuals. I suppose that with classical methods, representative long-term estimates of individual feeding niches are most difficult in the case of non-territorial individuals such as adolescent eagles, nomadic floaters or migrating eagles from the northernmost part of the Palaearctic (Thiollay 1994).

(2) I additionally adopted a relatively novel approach in raptor diet studies – the analysis of stable isotopes in tissues of predators and their potential food sources. The finding of DeNiro and Epstein (1978) "you are what you eat" also applied to white-tailed eagles. The determination of stable isotope ratios allowed both identification and quantification of eagle food sources from different habitats and trophic levels which was not limited to territorial individuals. The comparison of the isotopic composition of tissues deposited at different times such as liver and muscle tissue allowed the identification of eagles that shifted diets over time and eagles where diets remained constant. Thus, stable isotope analysis enabled longitudinal estimates of foraging patterns and niche width. With the prior knowledge of potential food sources gathered with the classical approaches, it was possible to apply recently established Bayesian mixing models that provided a robust quantification of the major food types of white-tailed eagles. Especially with respect to ephemeral food remains such as gut piles, this approach was very worthwhile. In summary, the main advantages of the use of stable isotope analyses are (i) the possibility to obtain long-term data on the diet of both populations and individuals, and (ii) the robust assessment of the contribution of potential food sources to raptor diets. As the estimates of the proportion of specific food sources provided by model outputs were consistent with estimates based on classical approaches, this study demonstrated

that stable isotope mixing models exhibit high explanatory power in describing raptor diets. The main drawback is that this method only reveals detailed information on dietary composition if prior knowledge about potential food species is available.

I conclude that the analysis of stable isotopes is suitable for longitudinal records of dietary variation among and within raptor populations and individuals. The ease and simplicity by which isotopic data can be gathered provide a window into raptor foraging patterns that could hardly be gathered with conventional techniques. As an alternative to the investigation of tissues from dead individuals, the stable isotope analysis of feathers, which reflect diet during their growth phase (Hobson & Clark 1992; Martínez del Rio *et al.* 2009), could constitute a non-invasive approach to examine the feeding ecology in raptors. I suggest that the ideal method to get detailed information on dietary composition comprises a combination of classical approaches as the first step, followed by the use of stable isotope mixing models to quantify and verify the results as a second step.

Foraging strategies of white-tailed eagles – key influences and optimality considerations

From an ecological point of view, this thesis provided new insights into the foraging strategies of white-tailed eagles. Considering its high diet diversity, this raptor, like many other raptor species, was previously regarded as a generalist and opportunist (Oehme 1975; Helander 1983; Sulkava *et al.* 1997). My results document that this is not the full picture:

Although white-tailed eagles can be characterised as generalist predators on a population level, individuals cannot be treated as ecologically equivalent in terms of their niche breadth, since I documented both dietary generalisation and specialisation in individual white-tailed eagles (Chapters 2 and 4). This finding is consistent with the growing evidence for individual differences in diet and foraging behaviour across a wide range of species and ecosystems (Angerbjörn *et al.* 1994; Bolnick *et al.* 2003; Newsome *et al.* 2009). Indeed, the foraging strategy of white-tailed eagles is very flexible – not as a result of simple opportunistic behaviour but because of complex decisions made in response to several factors (Fig. 6.1). I identified the availability of food resources as the most important factor that underlies the foraging strategy of white-tailed eagles, which corresponds to widespread empirical evidence that food availability is a key driver of a species' ecology (e.g., Brommer *et al.* 2002; Rutz & Bijlsma 2006; Millon *et al.* 2009). Food availability influences the foraging strategy of white-tailed eagles both directly and indirectly owing to its impact on other crucial factors

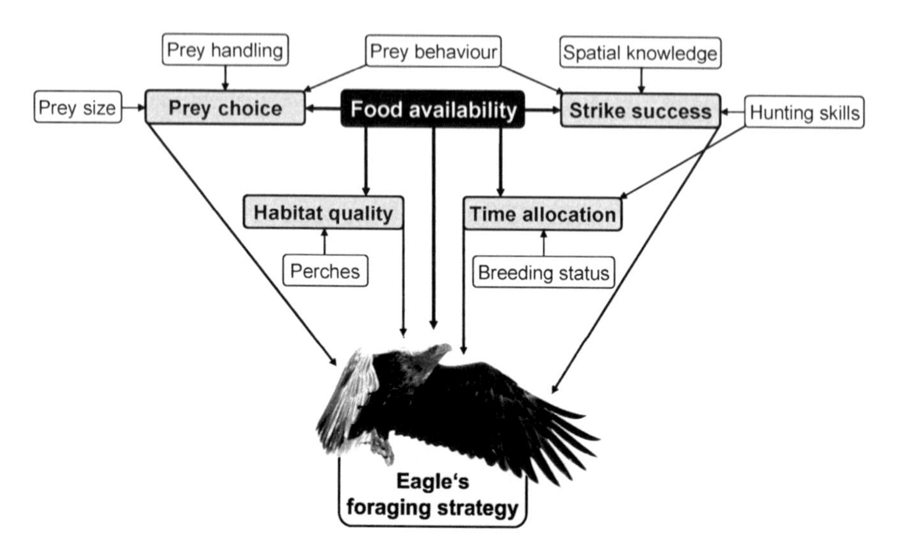

Fig.6.1. *Key factors identified in the present study that influence the foraging strategy of white-tailed eagles. The factors are illustrated in three hierarchy levels. The primary factor is represented by the black box; secondary and tertiary factors are depicted by grey and white boxes, respectively.*

that affect eagle foraging behaviour, such as habitat quality (see Chapters 2, 3 and 4), time allocation (see Chapter 3), prey choice (see Chapters 2 and 3) and strike success (see Chapter 3; Fig. 6.1). Habitat quality shapes eagle foraging with regard to the availability of profitable hunting grounds and suitable perches. Time allocation depends on individual hunting skills and breeding status and determines whether eagles pursue a sit-and-wait or active hunting mode. In contrast to opportunistic predators that consume food according to its abundance and distribution in the environment (Trayler *et al.* 1989; Hall-Aspland *et al.* 2005; Watters 2010), eagles are selective predators that rank their prey based on a consideration of several factors such as search and handling time, body size and anti-predator behaviour rather than only the availability of potential food sources. Finally, the strike success, which varies with eagle hunting skills and spatial knowledge as well as anti-predator behaviour of the target prey, affects the foraging strategy of white-tailed eagles.

It remains to be clarified whether the foraging strategy of individual white-tailed eagles can be regarded as "optimal" in terms of optimal foraging theory (Schoener 1971). Predictions of optimality models have often been supported in experimental approaches or relatively simple predator-prey systems (e.g., Werner & Hall 1974; Krebs *et al.* 1977; Leclerc 1991; Costa *et al.* 2008). However, the foraging strategy of white-tailed eagles consists of a complex interaction between external factors and foraging decisions in fluctuating environments. Are white-tailed eagles able to continually update their estimates of capture rates and availability to make decisions that maximise their efficiency in harvesting food?

When profitable food was abundant in areas accessible for white-tailed eagles, this species selected its diet both among and within prey classes. Among prey classes, eagles preferred the consumption of fish. Especially during the spawning season in spring, when fish was very abundant close to the water surface in the littoral zone, all study eagles focused on this food type which at this point was highly profitable in terms of search and handling time. Within prey classes, eagles chose species that provide benefits in terms of energy content and hunting effort as they preferred large over small fish and clumsy over agile waterfowl species. Consequently, our data support the first prediction of the optimal diet model that predators should prefer the most profitable food type if it is "sufficiently" available (Pyke *et al.* 1977).

Furthermore, this study documented a functional response of white-tailed eagles to changes in food availability on the basis of prey classes. When the availability of their primary food type fish declined, eagles included other food types such as waterfowl and mammalian carrion in their diet. The consumption of carrion, the food type with the lowest profitability in terms of search and handling time, was only related to the availability of fish and not to its own availability. The consumption of waterfowl, more profitable than carrion but less profitable than fish, was both related to fish availability and its own availability. These results correspond to the second prediction of the optimal diet model, namely that the inclusion of other food types in the diet should depend not on their own abundance but on the abundance of more profitable food types (Pyke *et al.* 1977).

The dietary niche of white-tailed eagles expanded, both within and between individuals, when seasonal fluctuations or habitat heterogeneity caused a reduction in the availability of profitable food types. I found this increase in diet diversity not only among but also within prey classes. This finding is consistent with the third prediction of the optimal diet model that a decline in food availability should result in a wider diet niche (Pyke *et al.* 1977).

In conclusion, the present study provides for the first time comprehensive evidence for food selection in a large raptor species such as the white-tailed eagle. I demonstrated that the large population dietary niche can include a considerably narrower individual niche width and is not equivalent to opportunistic foraging behaviour. Rather, the individual foraging strategy of white-tailed eagles is a product of adjustments to several individual and environmental factors that result in an efficient food intake in terms of optimal foraging theory.

Implications for white-tailed eagle conservation with respect to lead poisoning

In view of the conservation issues, this dissertation was supposed to contribute to an improvement of the conservation management of white-tailed eagles with respect to lead poisoning. I provided answers for the following two open questions posed by the relevant stakeholders (i.e., representatives of hunting associations, the ammunition industry, foresters, conservationists, veterinarians and wildlife biologists) concerning (1) the sources of lead poisoning, and (2) a possible solution to solve this acute problem in white-tailed eagles:

(1) What are the main sources of lead poisoning in white-tailed eagles?

The classical techniques for studying raptor diets give empirical evidence that carcasses of shot mammals and their gut piles, suggested to be the main lead sources in avian scavengers (Kim *et al.* 1999; Meretsky *et al.* 2000; Hunt *et al.* 2006), constitute an important alternative food source for the white-tailed eagle in Germany (Chapter 2). This finding is supported by a robust quantitative assessment of the contribution of shot mammalian carrion to eagle diet via stable isotope analysis (Chapter 4). Both classical and stable isotope approaches revealed that the dietary shift towards game mammal carrion occurs in autumn and winter months, when the incidence of lead poisoning in eagles is drastically increased (Krone *et al.* 2009). Consistent with this, I found in the stomachs of lead-poisoned eagles predominantly remains of large game mammals such as deer and wild boar. Also consistent with these findings, the video surveillance of game mammal carcasses at typical hunting glades during the main hunting season showed that white-tailed eagles are one of the most important resident scavenger species (Chapter 5). In view of these matching results, I draw the conclusion that hunter-shot mammals contaminated with lead fragments are the main source of lead poisoning in white-tailed eagles in Germany.

(2) Why were only small bullet fragments found in gizzards of white-tailed eagles? Do eagles avoid the ingestion of large metal fragments?

The feeding experiments described in Chapter 5 suggested that the gizzards of lead-poisoned white-tailed eagles contain only small bullet fragments because this species is in fact capable to avoid the ingestion of large metal fragments. Observations in the field indicated that not only white-tailed eagles but also other avian scavengers such as ravens (*Corvus corax*) and buzzards (*Buteo buteo*) avoid the intake of large metal particles during food processing. As this ability increased by all test animals with particle size, and non-toxic iron particles of 8.8 mm diameter were almost completely avoided, bullets that fragment into particles with a minimum size that exceeds this threshold may prevent the risk of metal ingestion by avian scavengers. Currently, there are already several lead-free bullet types, usually made of copper or copper alloys such as brass and less toxic for birds than lead (Kenntner *et al.* 2008), which could be considered by this criterion to be appropriate for game mammal hunting (Rosenberger 2007; Trinogga *et al.* 2008). Even if the use of such lead-free bullets protected avian scavengers from metal ingestion and poisoning, it has to be tested whether such ammunition is acceptable in hunting practise. Lead-based rifle bullets have been used by hunters for more than one hundred years and are well known for their killing efficacy. When striking body tissue, the high degree of fragmentation of lead-based rifle bullets into multiple, predominantly small particles has been documented to cause massive wounds because of increased tissue disruption and therefore a quick death (Liu *et al.* 1982; Fackler *et al.* 1984). If lead-free bullets only deform or fragment into a few large fragments, it is unclear whether this might reduce the efficacy to kill the target animal, and then create a welfare issue. Of course, the use of bullets that cause unnecessary pain for shot game mammals would not really be a sustainable solution for the lead problem. To address this issue, my colleague Anna Trinogga compared the killing efficacy of several types of lead-based and lead-free rifle bullets. As she found no significant differences in the killing efficacy between lead-based and lead-free rifle bullets, lead-free ones may be regarded as equivalent to lead-based ones in terms of their animal welfare effect (A. Trinogga *et al.* unpublished data). Consequently, I feel confident that the use of appropriate lead-free rifle bullets which fulfil the criteria identified in the feeding experiments in this study is a suitable approach to solve the problem of lead poisoning without currently known side effects.

Implications for future research on white-tailed eagles

The main data generated for this dissertation were derived from territorial white-tailed eagles which inhabited home ranges in a high density population in a highly suitable habitat (Fig. 1.2; Scholz 2010). Therefore, the next step in studying the feeding ecology of white-tailed eagles could be a focus on individuals in habitats of lower quality. As I identified the food availability as the key factor for shaping foraging strategies in white-tailed eagles, habitats without large water bodies with a high abundance of preferred fish or waterfowl species might incite other and possibly opportunistic foraging strategies. This could result in a higher proportion of game mammal carrion in eagle diets, as shown by the study pair SP7 in the present study, and thus in increased lead exposure.

The results on the foraging strategies of white-tailed eagles are quite consistent with the predictions of optimal diet models. However, additional research is needed to determine whether preferred prey species are an optimal choice for white-tailed eagles, as this study did not include factors such as digestibility or nutrient quality of potential food items, which may also be important for the assessment of the optimality of foraging (Krebs 1978). Furthermore, an experimental approach to study optimal decision rules in white-tailed eagles could be developed on the basis of my field observations to quantitatively define the limit up to which an increase in fish size corresponds to an increase in profitability.

This study was a first step to investigate the handling of food by raptors in general and of white-tailed eagles in particular with a focus on the issue of metal ingestion and lead poisoning. A future step could be the application of the outlined experimental approach to study in detail the handling of food by other scavenger species exposed to lead. Especially for species of international conservation concern, such as the vulnerable Steller's sea eagle (*Haliaeetus pelagicus*), the endangered Spanish imperial eagle (*Aquila adalberti*), or the critically endangered California condor (*Gymnogyps californianus*) as well as the white-rumped vulture (*Gyps bengalensis*; Fisher *et al.* 2006), this would be a worthwhile approach to improve conservation management and provide a sustainable solution to reduce the risk of exposure to lead.

References

Angerbjörn, A., Hersteinsson, P., Lidén, K. & Nelson, E. 1994. Dietary variation in arctic foxes (*Alopex lagopus*) – an analysis of stable carbon isotopes. *Oecologia* 99: 226-232.

Bolnick, D.I., Svanbäck, R., Fordyce, J.A., Yang, L.H., Davis, J.M., Hulsey, C.D. & Forister, M.L. 2003. The ecology of individuals: incidence and implications of individual specialization. *American Naturalist* 161: 1-28.

Brommer, J.E., Pietiäinen, H. & Kolunen, H. 2002. Reproduction and survival in a variable environment: ural owls (*Strix uralensis*) and the three-year vole cycle. *Auk* 119: 544-550.

Costa, G.C., Vitt, L.J., Pianka, E.R., Mesquita, D.O. & Colli, G.R. 2008. Optimal foraging constrains macroecological patterns: body size and dietary niche breadth in lizards. *Global Ecology and Biography* 17: 670-677.

DeNiro, M.J. & Epstein, S. 1978. Influence of diet on the distribution of carbon isotopes in animals. *Geochimica et Cosmochimica Acta* 42: 495-506.

Fackler, M.L., Surinchak, J.S., Malinowski, J.A. & Bowen, R.E. 1984. Bullet fragmentation: a major cause of tissue disruption. *Journal of Trauma* 24: 35-39.

Fisher, I.J., Pain, D.J. & Thomas, V.G. 2006. A review of lead poisoning from ammunition sources in terrestrial birds. *Biological Conservation* 131: 421-432.

Hall-Aspland, S.A., Hall, A.P. & Rogers, T.L. 2005. A new approach to the solution of the linear mixing model for a single isotope: application to the case of an opportunistic predator. *Oecologia* 143: 143-147.

Helander, B. 1983. Reproduction of the white-tailed sea eagle *Haliaeetus albicilla* (L.) in Sweden, in relation to food and residue levels of organochlorine and mercury compounds in the eggs. *Ph.D. dissertation*, University of Stockholm, Stockholm, Sweden.

Hobson, K.A. & Clark, R.G. 1992. Assessing avian diets using stable isotopes I: turnover of ^{13}C in tissues. *Condor* 94: 181-188.

Hunt, W.G., Burnham, W., Parish, C.N., Burnham, B., Mutch, B. & Oaks, J.L. 2006. Bullet fragments in deer remains: implications for lead exposure in scavengers. *Wildlife Society Bulletin* 34: 167-170.

Kenntner, N., Dänicke, S., Szentiks, C. & Krone, O. 2008. Vorläufige Ergebnisse – Toxizität alternativer Geschossmaterialien im Vogelmodell. In: *Bleivergiftungen bei Seeadlern: Ursachen und Lösungsansätze. Anforderungen an bleifreie Büchsengeschosse*, (ed.) Krone, O., pp. 21-30. Leibniz Institute for Zoo and Wildlife Research, Berlin, Germany.

Kim, E.Y., Goto, R., Iwata, H., Masuda, Y., Tanabe, S. & Fujita, S. 1999. Preliminary survey of lead poisoning of Steller's sea eagle (*Haliaeetus pelagicus*) and white-tailed sea eagle (*Haliaeetus albicilla*) in Hokkaido, Japan. *Environmental Toxicology and Chemistry* 18: 448-451.

Krebs, J.R. 1978. Optimal foraging: decision rules for predators. In: *Behavioral ecology: an evolutionary approach*, (eds.) Krebs, J.R. & Davies, N.B., pp. 23-63. Blackwell Scientific Publications, Oxford, UK.

Krebs, J.R., Erichsen, J.T., Webber, M.I. & Charnov, E.L. 1977. Optimal prey selection in the great tit (*Parus major*). *Animal Behaviour* 25: 30-38.

Krone, O., Kenntner, N. & Tataruch, F. 2009. Gefährdungsursachen des Seeadlers (*Haliaeetus albicilla* L. 1758). *Denisia* 27: 139-146.

Leclerc, J. 1991. Optimal foraging strategy of the sheet-web spider *Lepthyphantes flavipes* under perturbation. *Ecology* 72: 1267-1272.

Liu, Y.Q., Wu, B.J., Xie, G.P., Chen, Z.C., Tang, C.G. & Wang, Z.G. 1982. Wounding effects of two types of bullets on soft tissue of dogs. *Acta Chirurgica Scandinavica, Supplementum* 508: 211-221.

Martínez del Rio, C., Sabat, P., Anderson-Sprecher, R. & Gonzalez, S.P. 2009. Dietary and isotopic specialization: the isotopic niche of three *Cinclodes* ovenbirds. *Oecologia* 161: 149-159.

Meretsky, V.J., Snyder, N.F.R., Beissinger, S.R., Clendenen, D.A. & Wiley, J.W. 2000. Demography of the California condor: implications for reestablishment. *Conservation Biology* 14: 957-967.

Mersmann, T.J., Buehler, D.A., Fraser, J.D. & Seegar, J.K.D. 1992. Assessing bias in studies of bald eagle food habits. *Journal of Wildlife Management* 56: 73-78.

Millon, A., Nielsen, J.T., Bretagnolle, V. & Møller, A.P. 2009. Predator-prey relationships in a changing environment: the case of the sparrowhawk and its avian prey community in a rural area. *Journal of Animal Ecology* 78: 1086-1095.

Newsome, S.D., Tinker, M.T., Monson, D.H., Oftedal, O.T., Ralls, K., Staedler, M.M., Fogel, M.L. & Estes, J.A. 2009. Using stable isotopes to investigate individual diet specialization in California sea otters (*Enhydra lutris nereis*). *Ecology* 90: 961-974.

Oehme, G. 1975. Ernährungsökologie des Seeadlers, *Haliaeetus albicilla* (L.), unter besonderer Berücksichtigung der Population in den drei Nordbezirken der Deutschen Demokratischen Republik. *Doctoral dissertation*, Universität Greifswald, Greifswald, Germany.

Pyke, G.H., Pulliam, H.R. & Charnov, E.L. 1977. Optimal foraging: a selective review of theory and tests. *The Quarterly Review of Biology* 52: 137-154.

Redpath, S.M., Clarke, R., Madders, M. & Thirgood, S.J. 2001. Assessing raptor diet: comparing pellets, prey remains, and observational data at hen harrier nests. *Condor* 103: 184-188.

Rosenberger, M.R. 2007. *Jagdgeschosse*. Motorbuchverlag, Stuttgart, Germany.

Rutz, C. & Bijlsma, R.G. 2006. Food-limitation in a generalist predator. *Proceedings of the Royal Society B* 273: 2069-2076.

Schoener, T.W. 1971. Theory of feeding strategies. *Annual Review of Ecology and Systematics* 2: 369-404.

Scholz, F. 2010. Spatial use and habitat selection of white-tailed eagles (*Haliaeetus albicilla*) in Germany. *Doctoral dissertation*, Freie Universität Berlin, Berlin, Germany.

Simmons, R.E., Avery, D.M. & Avery, G. 1991. Biases in diets determined from pellets and remains: correction factors for a mammal and bird-eating raptor. *Journal of Raptor Research* 25: 63-67.

Stephens, D.W. & Krebs, J.R. 1986. *Foraging theory*. Princeton University Press, Princeton, USA.

Sulkava, S., Tornberg, R. & Koivusaari, J. 1997. Diet of the white-tailed eagle *Haliaeetus albicilla* in Finland. *Ornis Fennica* 74: 65-78.

Thiollay, J.M. 1994. Family Accipitridae (hawks and eagles). In: *Handbook of the birds of the world: New World vultures to guineafowl*. Volume 2, (eds.) del Hoyo, J., Elliot, A. & Sargatal, J., pp. 52-205. Lynx Edicions, Barcelona, Spain.

Tjernberg, M. 1981. Diet of the golden eagle *Aquila chrysaetos* during the breeding season in Sweden. *Ecography* 4: 12-19.

Trayler, K.M., Brothers, D.J., Wooller, R.D. & Potter, I.C. 1989. Opportunistic foraging by three species of cormorants in an Australian estuary. *Journal of Zoology* 218: 87-98.

Trinogga, A., Jeuken, P., Kinsky, H., Walter, M. & Krone, O. 2008. Wirksamkeit und Masseverlust ausgewählter bleifreier und bleihaltiger Büchsen-Projektile beim Beschuss von ballistischer Seife. In: *Bleivergiftungen bei Seeadlern: Ursachen und Lösungsansätze. Anforderungen an bleifreie Büchsengeschosse*, (ed.) Krone, O., pp. 44-57. Leibniz Institute for Zoo and Wildlife Research, Berlin, Germany.

Watters, J.L. 2010. A test of optimal foraging and the effects of predator experience in the lizards *Sceloporus jarrovii* and *Sceloporus virgatus*. *Behaviour* 147: 933-951.

Werner, E.E. & Hall, D.J. 1974. Optimal foraging and the size selection of prey by the bluegill sunfish (*Lepomis macrochirus*). *Ecology* 55: 1042-1052.

SUMMARY

Although the feeding ecology of raptors has attracted the attention of many scientists, aspects such as niche breadth, foraging strategy and diet selection are still poorly understood. This is particularly true for large species such as the white-tailed eagle *Haliaeetus albicilla*. Representative assessments of the contribution of food sources containing toxins such as lead to raptor diets are scarce as they constitute a methodological challenge, but are essential for optimising conservation efforts. This dissertation aimed to provide comprehensive information on the feeding ecology of white-tailed eagles to shed light on the interaction of such top predators with their environment and to improve the conservation management of this species which is in the process of recovery after severe population declines.

Raptors such as white-tailed eagles that consume a wide array of prey are usually regarded as opportunistic and generalist predators. However, the results of this thesis modify this assumption by presenting white-tailed eagles as selective foragers with varying diet diversity. Classical dietary studies on seven eagle pairs from northeastern Germany revealed individual adjustment of niche breadth and foraging strategy to available food and habitat conditions by maximising profitability in terms of search and handling time, energy content and anti-predator behaviour of their prey. Consistent with these findings, stable isotope ($\delta^{13}C$, $\delta^{15}N$) analyses of tissues from German (N = 75), Finnish (N = 10), and Greenlandic (N = 10) white-tailed eagles indicated that the population niche width is mainly determined by individual variations. Local feeding niches differed with local food supplies: while German and Finnish eagles showed individual generalisation caused by temporal dietary shifts, Greenlandic eagles showed specialisation because of constant foraging patterns (**Chapters 2**, **3** and **4**).

Since raptors are often difficult to observe owing to cryptic habits and large foraging areas, little is known about their time allocation and hunting mode. This study demonstrates that white-tailed eagles expended most of their time to perching, which implies an energy maximising rather than time minimising foraging strategy. Since perch-hunting was more efficient than flight-hunting, “sit-and-wait” for prey generally seems to be a low-cost, highly profitable foraging mode. Nevertheless, white-tailed eagles are flexible in their foraging behaviour. Linear mixed models revealed a positive relationship between changing weather conditions, decreasing food availabilities or age-dependent marginal hunting experience and active food search (**Chapter 3**).

Lead poisoning affects numerous raptors and is the major cause of death in white-tailed eagles. Primary reason for intoxication is suggested to be lead fragments ingested while feeding on mammalian carcasses killed by lead-based bullets. This dissertation confirms this assumption. Multiple regression models indicated that the specific dietary response of white-tailed eagles to changing food availability or poor habitat quality leads to scavenging on lead-contaminated carrion. An increase in the consumption of mammalian carrion by eagles was positively correlated with an increase in the incidence of lead-poisoned eagles. Stomachs of lead-poisoned eagles predominantly contained ungulate remains. These findings are supported by the isotopic data from the tissues of German white-tailed eagles and 16 of their potential food species (N = 90). Mass-balance and Bayesian isotope mixing models provided robust quantitative evidences that game mammal carrion constitutes an important alternative diet component for white-tailed eagles during the hunting season (**Chapters 2** and **4**).

An experimental approach of this study to identify lead-exposed species in Germany and to provide a solution for lead poisoning revealed the following: Primarily avian scavengers such as white-tailed eagles, ravens and buzzards exploited mammalian carcasses at typical hunting glades and are thus exposed to lead bullet fragments. The preparation of experimental carcasses with non-toxic metal fragments of different sizes showed that at least eagles and probably also ravens and buzzards avoided the ingestion of large metal fragments during carcass processing. The percentage of metal fragments successfully avoided increased with fragment size. Fragments of 8.8 mm diameter were almost completely avoided. This finding indicates that the use of bullets fragmenting into particles greater than 9 mm when striking game mammal tissue, such as numerous lead-free bullets, may prevent metal ingestion and poisoning in avian scavengers (**Chapter 5**).

Overall, this dissertation showed that studying raptor diets requires the combination of classical methods with stable isotope chemistry for representative and comprehensive information. The foraging strategy of top predators such as the white-tailed eagle is influenced by both individual and environmental factors and corresponds to an efficient food intake in terms of optimal foraging theory. The main sources of lead fragments that induce fatal lead poisoning are shot mammalian carcasses. One approach to solve the lead poisoning problem can be the use of lead-free bullets by hunters that deform or fragment into sufficiently large particles which would be avoided by scavenging eagles and other birds exhibiting similar feeding behaviour.

ZUSAMMENFASSUNG

Obwohl die Nahrungsökologie von Greifvögeln das Interesse vieler Wissenschaftler auf sich zieht, sind Aspekte wie Nischenbreite, Nahrungssuchstrategie und Nahrungswahl noch wenig erforscht. Dies betrifft insbesondere große Arten wie den Seeadler *Haliaeetus albicilla*. Repräsentative Einschätzungen der Relevanz von Nahrungsquellen, die Giftstoffe wie Blei enthalten, sind selten. Solche Einschätzungen sind wichtig für die Optimierung von Artenschutzmaßnahmen, stellen aber eine methodische Herausforderung dar. Das Ziel dieser Arbeit war die umfassende Untersuchung der Nahrungsökologie von Seeadlern, um Aufschluss über das Zusammenspiel solcher Top-Prädatoren mit ihrer Umwelt zu geben und zum Schutz dieser Art beizutragen, die sich nach erheblichen Populationsrückgängen in einer Erholungsphase befindet.

Greifvögel mit breitem Nahrungsspektrum wie Seeadler werden gewöhnlich als opportunistische und generalistische Prädatoren angesehen. Die Ergebnisse dieser Arbeit modifizieren jedoch diese Annahme und zeigen, dass Seeadler selektive Nahrungssucher mit variierender Nahrungsdiversität sind. Klassische Untersuchungen an sieben Seeadlerpaaren aus Norddeutschland ergaben individuelle Anpassungen der Nischenbreite und Nahrungswahl an aktuelle Nahrungs- und Habitatbedingungen, um den Ertragswert der Nahrung in Bezug auf Such- und Behandlungszeit, Jagdaufwand und Energiegehalt zu maximieren. Übereinstimmend mit diesen Ergebnissen zeigten Analysen stabiler Stickstoff- und Kohlenstoffisotope von Geweben deutscher (N = 75), finnischer (N = 10) und grönländischer (N = 10) Seeadler, dass die Nischenbreite einer Population hauptsächlich durch individuelle Variationen bestimmt wird. Lokale Nahrungsnischen variierten mit lokalem Nahrungsangebot: Während deutsche und finnische Seeadler individuelle Generalisierung zeigten, verursacht durch zeitliche Verschiebungen im Nahrungsspektrum, waren grönländische Seeadler Nahrungsspezialisten mit konstanten Nahrungssuchmustern (**Kapitel 2**, **3** und **4**).

Da viele Greifvögel aufgrund versteckter Lebensweise und großer Streifgebiete schwer zu beobachten sind, ist über ihre Jagdaktivität und ihr Jagdverhalten wenig bekannt. Diese Studie demonstriert, dass Seeadler die meiste Zeit für die Ansitzjagd aufwenden, was eher eine Nahrungssuchstrategie zur Energiemaximierung als zur Zeitminimierung impliziert. Da diese passive Nahrungssuche effizienter war als aktive Suchflüge, scheint "Sitzen-und-Warten" auf Beute eine mit geringen Kosten und hohem Nutzen verbundene Jagdmethode zu sein.

Trotzdem sind Seeadler flexibel in der Strategie ihrer Nahrungssuche. Lineare gemischte Modelle ergaben, dass sie auf veränderte klimatische Verhältnisse, eingeschränkte Nahrungsverfügbarkeit oder altersbedingt marginale Jagderfahrung mit gesteigerter aktiver Nahrungssuche reagieren (**Kapitel 3**).

Bleivergiftungen beinträchtigen zahlreiche Greifvögel und sind eine der häufigsten Todesursachen bei Seeadlern. Die vermutete Hauptquelle für Bleivergiftungen sind Bleipartikel, aufgenommen während des Fressens an Säugetierkadavern, welche mit bleihaltiger Jagdmunition erlegt wurden. Diese Arbeit bestätigt diese Annahme. Multiple Regressionsmodelle zeigten, dass die funktionale Antwort von Seeadlern auf veränderte Nahrungsverfügbarkeit oder geringe Habitatqualität zur gesteigerten Aufnahme von bleikontaminiertem Aas führt. Ein Anstieg im Verzehr von Säugetierkadavern korrelierte positiv mit einem Anstieg von Bleivergiftungen bei Seeadlern. Die Mägen bleivergifteter Seeadler enthielten fast ausschließlich Reste von Paarhufern. Diese Ergebnisse sind durch die Isotopenanalysen der Gewebe deutscher Seeadler und 16 ihrer potentiellen Nahrungstierarten ($N = 90$) gestützt. Massenbilanz und Bayessche Isotopen-Mischmodelle lieferten robuste quantitative Beweise, dass Kadaver von jagdbaren Säugetieren eine wichtige alternative Nahrungskomponente für Seeadler während der Jagdsaison darstellen (**Kapitel 2** und **4**).

Ein experimenteller Ansatz dieser Studie zur Identifizierung bleiexponierter Arten in Deutschland und zur Entwicklung einer Lösungsmöglichkeit für die Bleiproblematik ergab Folgendes: Vorwiegend aasfressende Vögel wie Seeadler, Kolkraben oder Mäusebussarde nutzten Säugetierkadaver an Jagdlichtungen als Nahrungsquellen und sind somit Bleigeschosspartikeln ausgesetzt. Die Präparierung von Versuchskadavern mit nicht-toxischen Metallpartikeln unterschiedlicher Größe zeigte, dass mindestens Seeadler und wahrscheinlich auch Raben und Bussarde große Metallpartikel bei der Nahrungsaufnahme vermieden. Der Prozentsatz erfolgreich vermiedener Metallpartikel stieg mit der Partikelgröße. Partikel mit einem Durchmesser von 8.8 mm wurden nahezu vollständig vermieden. Diese Erkenntnis weist darauf hin, dass eine Verwendung von Jagdgeschossen, welche sich beim Aufprall in Partikel mit einem Durchmesser größer als 9 mm zerlegen, wie zahlreiche bleifreie Geschosse, das Risiko für die Aufnahme von Geschossresten und einer daraus resultierenden Vergiftung bei aasfressenden Vögeln erheblich verringert (**Kapitel 5**).

Insgesamt ergab die vorliegende Arbeit, dass nahrungsökologische Studien an Greifvögeln für repräsentative und umfassende Informationen eine Kombination aus klassischen Methoden und moderner Isotopenanalyse erfordern. Die Nahrungssuchstrategie von Top-Prädatoren wie dem Seeadler ist sowohl durch individuelle als auch Umweltfaktoren beeinflusst und entspricht einem effizienten Nahrungserwerb im Sinne der Optimalitätstheorie. Die Hauptquellen für Bleivergiftungen sind Bleigeschosspartikel in den Geweben erlegter Säugetierkadaver. Ein Lösungsansatz für die Bleiproblematik ist der Einsatz bleifreier Munition, welche deformiert oder sich in ausreichend große Partikel zerlegt, die von aasfressenden Seeadlern und anderen Vögeln mit ähnlichem Fressverhalten vermieden werden.

DANKSAGUNG

Mein besonderer Dank gilt meinem Betreuer Dr. Oliver Krone. Er eröffnete mir den Weg in die angewandte Forschung und ermöglichte es mir, mich im Rahmen seines Projekts mit dieser interessanten Fragestellung zu befassen und gleichzeitig einen Beitrag zum Greifvogelschutz zu leisten. Darüber hinaus bin ich ihm dankbar für seine vielfältige Förderung und Unterstützung, die Bereitstellung umfangreichen Probenmaterials sowie die Herstellung wichtiger Kontakte.

Prof. Dr. Heribert Hofer danke ich herzlich für sein stetes Interesse an meiner Arbeit, inspirierende fachliche Diskussionen sowie hilfreiche Kommentare und Korrekturen zu allen Manuskripten. Weiterhin danke ich ihm für die freundliche Begutachtung dieser Dissertation.

Vielen Dank an PD Dr. Christian Voigt für seinen fachlichen Rat sowie nützliche Anregungen und Korrekturen bei der Verfassung des Manuskripts zur Isotopenanalyse.

Prof. Dr. Silke Kipper danke ich vielmals für die freundliche Begutachtung dieser Dissertation.

Dem Bundesministerium für Bildung und Forschung (BMBF) sowie dem Leibniz-Institut für Zoo- und Wildtierforschung (IZW) danke ich für die Finanzierung dieser Doktorarbeit.

Mein aufrichtiger Dank gilt Jörg Gast und allen Mitarbeitern der Naturparkverwaltung Nossentiner/Schwinzer Heide für ihre logistische Unterstützung und Naturschutzarbeit, die mir eine Datenaufnahme in diesem wunderschönen Vogelrückzugsgebiet ermöglicht hat. Den Horstbetreuern Dr. Wolfgang Mewes und Dr. Wolfgang Neubauer danke ich für ihre tatkräftige Hilfe und das entgegengebrachte Vertrauen. Dr. Neubauer danke ich zudem für die freundliche Bereitstellung detaillierter Daten bezüglich der Wasservogelverfügbarkeit am Krakower Obersee.

Ulrich Reeps danke ich herzlich für die Unterkunft in seiner wunderschön und einsam gelegenen Feldstation, die über zwei Jahre mein zweites Zuhause war. Ich fühlte mich dort unbeschreiblich wohl und die Freundlichkeit, mit der Herr Reeps mich bereits am ersten Tag willkommen hieß, wie auch sein wunderbarer Humor, werden mir unvergessen bleiben.

Die Jäger, Landwirte und Forstämter in der Nossentiner Heide haben mir gezeigt, wie gut Artenschutz funktionieren kann, wenn alle Interessengruppen zusammenarbeiten! Ob es um die Bereitstellung von Kameraplätzen, Fallwild oder Jagdstreckendaten ging – ich wurde stets freundlich unterstützt. Besonders danken möchte ich dafür Mathias Seltmann, Hans-Jörg

Martinez und Rüdiger Pawelke. Dieser Zusammenarbeit verdanke ich einen wichtigen Teil meiner Arbeit.

Gleichermaßen gilt den Fischern der Seen, welche ebenfalls von meinen Beobachtungsadlern befischt wurden, mein herzlichster Dank. Frank und Wolfgang Geibrasch sowie Bernhard Birkholz haben ohne zu Zögern den Mehraufwand in Kauf genommen, wenn ich mit an Bord war. Insbesondere während der Laichzeiten, in denen ich tausende von Fischen zu zählen und messen hatte, war ihre tatkräftige Hilfe ein Geschenk des Himmels.

Mike Wuttge bin ich sehr dankbar für seine uneingeschränkte Hilfsbereitschaft bei allen erdenklichen Problemen, die sich während meiner Feldaufenthalte ergaben.

Dem Museum für Naturkunde in Berlin sowie Jürgen Fiebig, Detlef Willborn, Dr. Helmut Winkler und Manfred Lüpke danke ich vielmals für die Unterstützung bei der taxonomischen Einordnung der Beutereste. Dr. Winkler danke ich darüber hinaus für hilfreiche Tipps und die Bereitstellung von Literatur für die Alters- und Längenbestimmung gesammelter Fischreste.

Dem Leibniz-Institut für Gewässerökologie und Binnenfischerei (IGB) und Titus Rohde danke ich für die freundliche Leihgabe eines Wasserqualitätsmessgeräts.

Paul Sömmer, Dr. Torsten Langgemach, Rainer Altenkamp und Dr. Bernd-Ulrich Meyburg danke ich herzlich für die Möglichkeit, Fütterungsversuche mit Seeadlern durchzuführen, die temporär an der Naturschutzstation Woblitz untergebracht waren.

Ganz besonders möchte ich den Mitgliedern meiner Arbeitsgruppe danken, die alle einen großen Anteil am Gelingen meiner Arbeit haben. Friederike Scholz verbrachte mit mir viele Feldaufenthalte und steht mir seitdem so nahe, als wäre sie ein Teil meiner Familie. Justine Sulawa war mir eine tolle Kollegin und Freundin in Berlin, die ich sehr vermisse, seit sie wieder in Frankreich lebt. Friederike und Justine standen mir immer mit Rat und Tat zur Seite und wir gingen gemeinsam durch alle Höhen und Tiefen – „Wir sind für immer die drei Musketiere“! Auf Anna Trinogga konnte ich mich immer verlassen und Norbert Kenntner kümmerte sich stets gerne um alle „seine Mädels“. Kirsi Blank und Katrin Totschek danke ich für ihre tatkräftige Unterstützung und die fröhliche Arbeitsatmosphäre. Wir waren ein einmaliges Seeadler-Team und ich danke euch für alles!

Norbert, Andre Laubner und Torsten Lauth danke ich für ihren engagierten und unermüdlichen Einsatz, bei jeder Wetterlage Seeadlerhorste für mich zu erklettern.

Jan Axtner, Niko Balkenhol, Sarah Benhaiem, Aines Castro Prieto, Simon Ghanem, Mirjam Gippert, Sabine Greiner, Deike Hesse, Ina Leinweber, Sebastian Menke, Yvonne

Meyer-Lucht, Zoltan Mezö, Kristin Mühldorfer, Kathleen Röllig, Pablo Santos, Nina Schwensow, Rahel Sollmann, Andreas Wilting und vielen anderen danke ich sehr für eine unvergessliche Zeit am IZW und in Berlin. Meinen ehemaligen Mitbewohnerinnen Fanny Rexa, Patricia Kunert und Hanne Böttcher danke ich für den tollen Start in Berlin sowie viele schöne gemeinsame Unternehmungen.

Ein großes Dankeschön an Maren Huck, Rahel, Deike und Niko für wertvolle Kommentare und Korrekturen, welche diese Arbeit sehr aufgewertet haben. Friederike und Stephanie Kramer-Schadt danke ich auch sehr für die Einführung und tatkräftige Unterstützung im Umgang mit ArcView und GIS, Stephanie zudem für ihre freundliche Hilfe bei allen Fragen rund um R.

Vielen Dank auch allen anderen Mitarbeitern des IZW für ihre Freundlichkeit und Unterstützung. Besonders danke ich Karin Sörgel für die Durchführung hunderter Isotopenmessungen sowie Dagmar Viertel für die Anfertigung unzähliger elektronenmikroskopischer Haaraufnahmen. Knut Eichhorn steckte seinen ganzen Erfindungsgeist in meine Videoüberwachungssysteme und Wolfgang Richter baute mir so manchen nützlichen Ausrüstungsgegenstand. Wolfgang Tauche war bei jedem Computerproblem sofort zur Stelle und Jürgen Streich hatte stets ein offenes Ohr für statistische Fragen. Beate Peters-Mergner und Cornelia Greulich beschafften mir immer in Rekordzeit jeden gewünschten Artikel.

Mein unermesslicher Dank gilt meiner Familie. Meine Eltern Bärbel und Massod Nadjafzadeh sowie meine Schwester Susanne waren immer für mich da und ermöglichten es mir, meinem angestrebten Lebensweg zu folgen. Mein Lebensgefährte Karsten Schwaak unterstützte mich darin, meine Träume zu realisieren, war stets sehr verständnisvoll und baute mich in Zeiten einer Schaffenskrise immer wieder auf. Vor allem in der Endphase haben sein positives Denken und seine Energie mir die notwendige Kraft gegeben. Seine Eltern Ursula und Manfred Schwaak sowie seine Schwester Simone unterstützten mich auf jede erdenkliche Weise. Ich empfinde es als ganz besonderes Glück, solch wundervolle Menschen um mich herum zu haben.

Zu guter Letzt möchte ich an dieser Stelle denjenigen danken, ohne die es diese Arbeit nicht gäbe – meinen Beobachtungsadlern, die im Fluge mein Herz eroberten. Ich hoffe sehr, ihre freundliche Zusammenarbeit damit zu belohnen, dass diese Arbeit dazu beiträgt, Bleivergiftungen und den damit verbundenen qualvollen Tod für Seeadler und andere Greifvögel zu verhindern!